Springer Proceedings in Physics 14

Springer Proceedings in Physics

Managing Editor: H. K. V. Lotsch

Springer Proceedings in Physics is a new series dedicated to the publication of conference proceedings. Each volume is produced on the basis of camera-ready manuscripts prepared by conference contributors. In this way, publication can be achieved very soon after the conference and costs are kept low; the quality of visual presentation is, nevertheless, very high. We believe that such a series is preferable to the method of publishing conference proceedings in journals, where the typesetting requires time and considerable expense, and results in a longer publication period. Springer Proceedings in Physics can be considered as a journal in every other way: it should be cited in publications of research papers as *Springer Proc. Phys.*, followed by the respective volume number, page number and year.

Magnetic Properties of Low-Dimensional Systems

Proceedings of an International Workshop
Taxco, Mexico, January 6–9, 1986

Editors: L. M. Falicov and J. L. Morán-López

With 126 Figures

Springer-Verlag Berlin Heidelberg New York
London Paris Tokyo

Professor Leopoldo M. Falicov
University of California, Berkeley, Dept. of Physics,
Berkeley, CA 94720, USA

Professor José L. Morán-López
Dept. de Física, CINVESTAV-IPN, Apdo. Postal 14-740,
07000 México, D.F., México

ISBN-13: 978-3-642-71014-8 e-ISBN-13: 978-3-642-71012-4
DOI: 10.1007/978-3-642-71012-4

Softcover reprint of the hardcover 1st edition 1986

2153/3150-543210

Preface

The International Workshop on the Magnetic Properties of Low-Dimensional Systems was held in the Hotel Posada de la Misión, Taxco, in the state of Guerrero, Mexico, January 6–9, 1986. It was part of the scientific activities organized to celebrate the 25th anniversary of the Centro de Investigación y de Estudios Avanzados, Instituto Politécnico Nacional, Mexico City. It was supported by the Consejo Nacional de Ciencia y Tecnología (Mexico), the Secretaría de Educación Pública (Mexico), the International Centre for Theoretical Physics (Trieste, Italy), the Departimento per la Cooperazione allo Sviluppo (Italy), the National Science Foundation (United States), and the British Council (United Kingdom).

The city of Taxco is a living monument to colonial Mexico. Its twisting, steep, and cobblestoned streets contain buildings that date as far back as 1534. It derives its name from "Tlachco", a Nahoa (local Indian) word for a pre-Columbian ball game – a well-preserved ball field can be seen and was visited by the conferees in Xochicalco, in the state of Morelos, an archeological site not far from Taxco. It was in these historic, beautiful and almost perfect surroundings that the International Workshop took place.

There were 67 participants from all over the world: 44 from Mexico, 11 from the United States, 4 from the Federal Republic of Germany, 2 each from France and Japan, and one each from Argentina, Brazil, England, and Spain. The meeting consisted of invited lectures divided into seven sessions (four morning and three evening). With the exception of the opening remarks (by Dr. Héctor O. Nava Jaimes, Director of the Centro de Investigación y de Estudios Avanzados, Instituto Politécnico Nacional) and the closing summary (by Professor L.M. Falicov, Chairman of the Conference), all presentations are included in this volume.

The theme of the meeting – magnetism in low-dimensional systems – has become a very active subject of truly interdisciplinary character, spanning such diverse areas as complex electronic-structure calculations; the chemistry and metallurgy of transition-metal and rare-earth compounds; the technology of thin films and overlayers (e.g., molecular beam epitaxy); surface science; the chemistry, electrical engineering, and materials science of small particles; the chemistry and physics of intercalation; Mössbauer, spin-resonance, optical, and electron spectroscopies; the interplay of magnetism with other condensed-matter phenomena (e.g., superconductivity and charge- and spin-density waves); and the technology of magnetic recording. Obviously an in-depth view of this vast field could not be given by 23 speakers in three and a half days. It is surprising and gratifying, however, to see how much of the field was covered and how many of the recent advances were reported and discussed. In that respect, this volume is a good representation of the state of the art in January 1986.

To put the various problems in perspective, it is useful to classify loosely the various systems under discussion here according to their restricted dimensionality. If a magnetic atom – a transition metal or a rare earth – is considered of dimension zero, and a uniform, crystalline, magnetic, bulk solid is the prototype of three dimensions, a chart of the systems discussed in this volume should look as follows:

DIMENSION	SYSTEM
0	Atoms
	Molecules
	Small clusters
	Clusters
	Small particles
1	Disordered chains
	Ideal chains
	Linear compounds
	Ridges on surfaces
2	Ideal two-dimensional structures
	Thin films
	Overlayers
	Surface layers
	Intercalated layered structures
	Layered structures
	Surface-bulk interactions
	Heterostructures (super-periodic)
	Disordered solids
3	Crystalline solids

The theoretical contributions to this volume, which explore most of the systems mentioned above, can be divided into four categories:

1. Microscopic quantum theories:

1A) Realistic | 1a) Schrödinger | 1α) Local density approximation

1B) Simple models | 1b) Dirac | 1β) Many-body

2. Microscopic classical theories:

2A) Statistical mechanics | 2a) Renormalization group

2b) Perturbation and series expansions

2c) Cluster variation method

3. Numerical simulations:

3A) Monte Carlo

3B) Molecular dynamics

4. Macrosocpic theories:

4A) Classical elasticity and electromagnetic theory

The experimental contributions are equally varied. In addition to the range of magnetic systems that have been explored being large, the variety of probes used increases as probes of higher intensity and better resolution become more easily available. There are essentially three kinds of probes: photons, electrons, and nuclear particles. They are preferred if they are polarized, in intense beams, and with monochromatic, tunable energies.

The experimental conditions – the so-called control parameters – are also fully developed; in particular, the following have been used or achieved:

a) Very low and very high temperatures;
b) Good surface preparation (high-vacuum techniques);
c) Good surface characterization techniques;
d) Very high and homogeneous magnetic fields;
e) Novel and sophisticated synthetic chemistry;
f) Sophisticated sample-growth techniques, e.g., molecular beam epitaxy.

And last but not least, the field derives its great vitality from the constant motion across boundaries of all sorts: between science and engineering, between basic research and applications, between theory and experiment, between academic research and industrial production, between the private and the public sectors, and across all international boundaries.

Berkeley, Mexico City
June 1986

L.M. Falicov
J.L. Morán-López

Contents

Part I

The Structure of Magnetic Surfaces

Surface Magnetism

G. Allan

Laboratoire des Surfaces et Interfaces, Institut Supérieur d'Electronique du Nord, 41, boulevard Vauban, F-59046 Lille Cedex, France

1. Introduction

Our knowledge of surface magnetism is constantly improving, but we are still at the beginning of this chapter of surface science. A lot of uncertainties remain, caused mainly by the difficulties in performing surface experiments. For example, different methods give many times different results for the same system. However, convergence of results now seems to be more common. From the theoretical point of view, the lack of periodicity perpendicular to the surface makes calculations much more difficult. In general a double approach is necessary. One must use simple models as well as more elaborate methods to study semi-infinite crystals or thin films.

In the next sections we shall only review a few experiments dealing with surface magnetism. This field is rapidly developing, but complete reviews have already been published elsewhere [1-4]. We shall mainly show that surface magnetism exists and discuss specific effects of the surface on magnetism. Finally, in the last section, we describe some applications to transition metals of the 3d series.

2. Dead Layers

The first observation of magnetically "dead" (i.e. paramagnetic) layers in thin Ni films on a Cu substrate was made fifteen years ago [5]. This concept remained current until very recently even though experimental results were not convincing and sometimes contradictory.

(a) Ferromagnetic Transition-Metal Thin-Film Magnetization

A transition-metal thin film is obtained by electrolytic deposition from an aqueous metal salt solution on a Cu substrate. Magnetization measurements made at room temperature are extrapolated to 0 K. This extrapolation shows that the first two deposited layers are nonmagnetic. As one increases the thin-film thickness, these two layers remain dead. The substrate (Cu or Au) does not seem to affect the results. This was interpreted as a surface effect. More recent experiments indicate that this interpretation is incorrect. Two other interpretations are plausible. The thin film is limited by two surfaces, or more precisely by two interfaces. One is the copper-nickel interface, the other is a nickel-electrolyte one. The latter is certainly covered by hydrogen adatoms and water molecules that can reduce the surface magnetic moment. This has been shown to be the case during hydrogen or hydrocarbon [6] adsorption. As we shall see below, recent experiments have also shown that the surface keeps a magnetic moment.

The anomalous Hall resistivity is able to trace magnetic moments in a metal; it has also been used near the surface [7]. In the surface case, a thin Ni film is deposited in ultra-high vacuum on a $Pb_{0.75}Bi_{0.25}$ substrate. The Ni film acts as a shunt for the $Pb_{0.75}Bi_{0.25}$ substrate, which is not superconducting at 10 K. By looking at the difference with the free substrate resistivity, one can measure the parallel Ni thin-film resistivity as a function of an applied magnetic field normal to its surface. Figure 1 schematically shows the results in the paramagnetic and

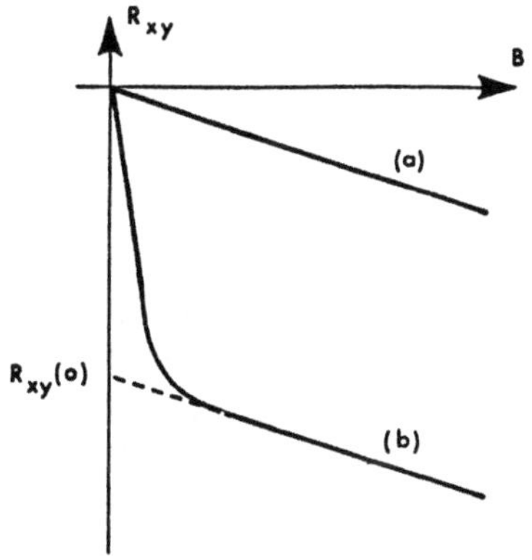

Figure 1. Hall resistivity as a function of the applied magnetic induction.(a) paramagnetic thin film; (b) ferromagnetic thin film.

ferromagnetic states. The resistivity R_{xy} is proportional to the sample magnetization. In the paramagnetic state, this magnetization is proportional to the applied field. In the ferromagnetic one, the resistivity strongly increases until saturation is reached. The extrapolation value $R_{xy}(0)$ from high fields is then proportional to the sample saturation magnetization, and dR_{xy}/dB is proportional to the sample susceptibility. The measurement shows that the susceptibility diverges for thicknesses between two and three adlayers [7]. For coverages below the one that gives this divergence, the magnetization is equal to zero. For thicker films the magnetization increases, rapidly reaching a bulk limiting value. An additional condensation of $Pb_{0.75}Bi_{0.25}$ on top of the Ni film decreases the magnetization by creating a new substrate-nickel interface. This seems to indicate that the dead layers occur near the substrate-nickel interface and not in the neighborhood of the free surface.

Some recent theoretical results for the Ni and Cu surfaces, and for the Cu-Ni interface [8,9] support these results. On a Cu(100) surface, a nickel monolayer is not dead. Its magnetic moment is close to the bulk value. Whereas on a Cu(111) surface a nickel monolayer is nearly paramagnetic. There is a competition between the d band narrowing for surface atoms and the sp-d hybridization which increases when a Ni atom has more noble-metal nearest neighbor atoms (as in $Ni_x Cu_{1-x}$). On a Ni(111) or (110) surface, the adsorption of copper reduces the magnetic moment. The magnetic moment strongly depends on the density of states at the Fermi level. For nickel, the Fermi level is close to the d band upper edge. Any d band narrowing or sp-d hybridization can affect the shape of the d band edge. On the other hand the iron Fermi level lies in a region of high density of states, and the corresponding effects are therefore less important. This would also explain the different behavior of Ni and Fe atoms near a Pd surface. Bulk iron impurities on Pd metal show giant moments (~15 Bohr magnetons), but bulk nickel impurities remain paramagnetic. Some time ago we discussed the surface enhancement of this moment for an iron surface impurity or adatom [10]. The local susceptibility, proportional to the local density of states at the Fermi level, increases as one goes from the bulk to the surface and to an adatom because of d band narrowing. Such an effect is also observed with Ni impurities, which are magnetic as adatoms but become paramagnetic impurities deeper in bulk Pd [11].

(b) Why Does Surface Magnetism Occur?

The results of Lieberman et al. have confused researchers for a long time. The difference between two-dimensional and surface magnetism was not clear.

It has been well known for a long time that two-dimensional (2D) surface magnetism does not exist [12,13]. The reasons are the same as the nonexistence of 2D lattices. At 0 K one can find a stable ordered state, but at any finite temperature the crystal excitations (magnons and phonons, respectively) destroy any long-range order. In a way similar to a three-dimensional crystal, the 2D magnetization may be written as [14]

$$\frac{M}{M_0} = 1 - \frac{1}{2\pi^2}\int_0^{\infty}\frac{kdk}{\exp\,(Jk^2/k_BT) - 1} \quad . \tag{1}$$

Similar to the phonon case, the low frequency limit induces a logarithmic divergence. But such a zero value is only reached for an infinite crystal. The lower limit of integration must be in fact replaced by a value of the order of (2π/Na) where N is the total number of atoms in the 2D crystal. For the phonon case, a 2D lattice is stable for temperatures less than (1000/LnN) K [15]. Typically N is of the order of 10^{16}. Such a crystal is stable for temperatures lower than 50 K. This is true for a pure 2D crystal with no substrate. In the case of a surface or an adlayer this is no longer true. The surface (or the adlayer) always interacts with the substrate. In some calculations one can often neglect the interaction I_s between the adlayer and the substrate compared to the direct interaction I between adatoms. In the phonon case, it has been shown that mean-square displacements (analogous to spin excitations) only diverge as Ln I/I_s. Since we can expect the interaction of the surface layer with the substrate to be of the same order of magnitude as the intra surface layer, such a divergence does not exist; the effect only enhances surface excitations because they are less energetic than the bulk ones. Let us for example consider the variation of the magnetic moment as one increases the temperature. No measurement has been made for 3d transition metals, except for Ni(100) near the Curie temperature. In this case, the critical behavior of the magnetization of the nickel top layer agrees with the results of an X-Y model. For T/T_c ranging from 0.9 to 0.992, the magnetization decreases with the critical exponent β equal to 0.825 [16]. Mössbauer experiments for free nickel surfaces have shown a magnetization that decreases faster than the bulk one. For the whole temperature range below the Curie point, spin polarized photoemission rate measurements have been done for Fe_3O_4 surfaces [17]. The surface magnetization (Figure 2) decreases faster than the bulk one, in agreement with theoretical predictions near the surfaces of Heisenberg ferromagnets [18]. So the 2D aspect of the surface only enhances spin excitations but does not destroy long-range order. It modifies the local density of states at the Fermi level and one may find some cases where the Stoner condition for magnetism is satisfied near the surface, although it is not satisfied in the bulk. We should then have "live" layers.

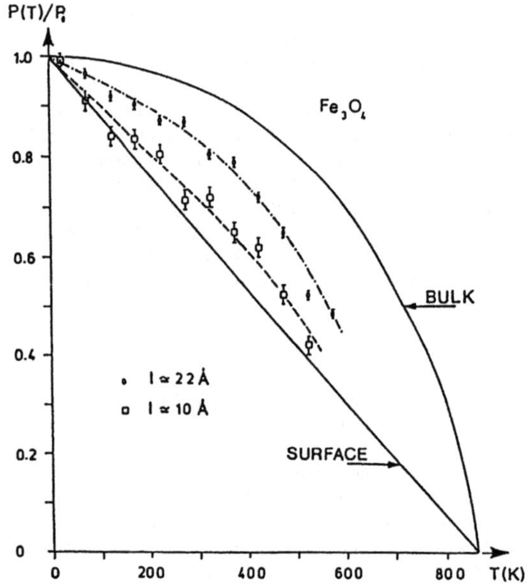

Figure 2. Fe_3O_4 spin-polarized photoemission rate for two photon energies 7.8 eV (ℓ = 22 Å) and 10.3 eV (ℓ = 10 Å) compared to the bulk magnetization and a linear surface variation.

3. Live Layers

We shall here discuss only the transition-metal free surfaces, mainly the Cr(100) surface. However, it seems that live layers may exist near the Si(111) surface [19] and near He_3-substrate interfaces [20].

(a) Surface Effects on the d Band

The coordination number of a surface atom is smaller than that of a bulk atom. Then there is a surface d-band narrowing [Figure 3(a)]. One then would expect a higher density of states near the middle of the series and a lower one near the edges. In fact, such a simple picture is modified by the screening effect of the d electrons. One has in fact to shift the surface d band in order to keep each layer almost neutral [21]. Then the surface local density of states is shifted upwards for a nearly filled band. So we do not expect large changes of density of states at the Fermi level, even for nickel. This very simple model agrees quite well with more complete d-electron tight-binding calculations. The surface magnetic moment is found to be close to the bulk one.

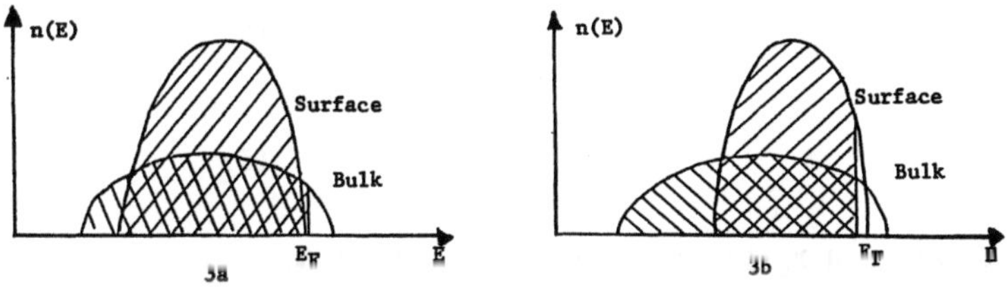

Figure 3. Schematic bulk and surface densities of states. (a) Non-selfconsistent result; (b) selfconsistent calculation.

Another surface effect is the surface bound states occurrence near the middle of the d band for body-centered-cubic transitions metals. Because of the small surface coordination number (4 instead of 8 in the bulk), a surface electron does not transfer easily onto its neighbors and remains a "long" time localized near the surface layer. So we get a surface resonance, and so a high density level in an energy range close to the Fermi level of the b.c.c. transition metals (Figure 4). This is certainly a major surface effect: because of the large increase of the density of states at the Fermi level, live layers may be induced. Let us now discuss a more accurate condition for the existence of a surface magnetic moment.

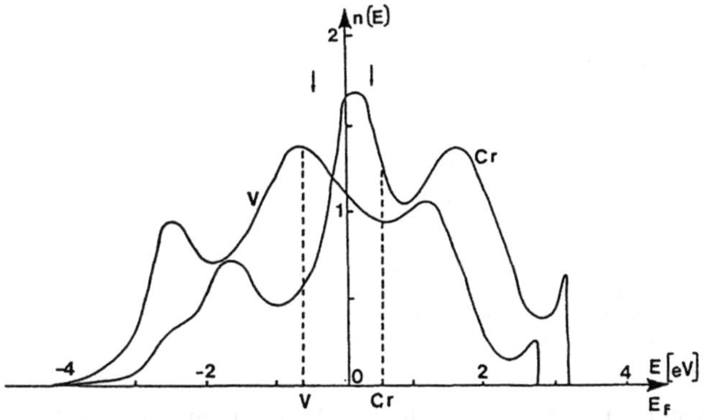

Figure 4. The (100) surface density of states of paramagnetic vanadium and chromium. Note the surface peaks close to the Fermi level.

(b) Condition for the Existence of a Surface Magnetic Moment [22]

In the tight-binding approximation, one can show that the criterion for the existence of live layers near the surface of a paramagnetic bulk metal is:

$$U_s(X_s + KX_I) > 1 , \qquad (2)$$

where U_s is the surface Coulomb integral, X_s and X_I are respectively the surface susceptibility and the susceptibility between two planes parallel to the surface. In fact, if one puts a potential V on all the surface atoms, in perturbation theory X_sV is the electron number change in the surface plane and X_IV in the plane just below the surface. The constant K gives the variation of the local magnetic moment M_i as one goes i layers inside the bulk

$$M_i = M_o K^i . \qquad (3)$$

In (3) M_o is the surface magnetic moment,

$$K = \frac{A \pm (A^2 - 4)^{1/2}}{2} , \qquad (4)$$

where

$$A = \frac{1 - 4X_B}{UX_I} , \qquad (5)$$

U is the bulk Coulomb integral, and X_B is the bulk equivalent of X_s.

For bulk magnetic materials, similar equations may be obtained. These equations describe the change in the magnetic moment near the surface. The results are given in Figures 5 and 6. Figure 6 shows that except near the limits of nonmagnetic and magnetic bulk states, the magnetic moment change near the surface decreases very rapidly (K is small). K is negative near the middle of the series due to an antiferromagnetic coupling between adjacent planes. Therefore the magnetic moment oscillates around its bulk value. Figure 5 shows that X_s is large near a half-filled band (even larger than the surface density). Due to the negative values of K and U_I in this energy region, the condition (2) is easily satisfied with the usual Coulomb integral values.

Let us examine the case of vanadium whose d-electron number N_d is close to 4. The surface Stoner criterion is satisfied for U_s larger than 0.5 eV. We can expect

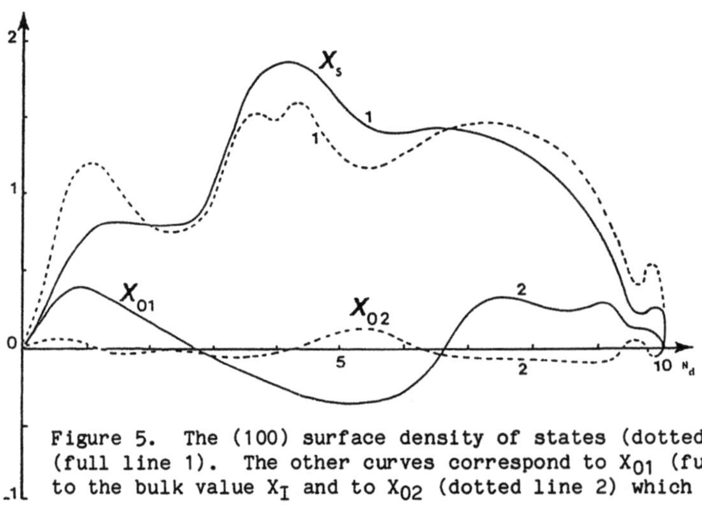

Figure 5. The (100) surface density of states (dotted line 1) compared with X_s (full line 1). The other curves correspond to X_{01} (full line 2) which is close to the bulk value X_I and to X_{02} (dotted line 2) which is quite small.

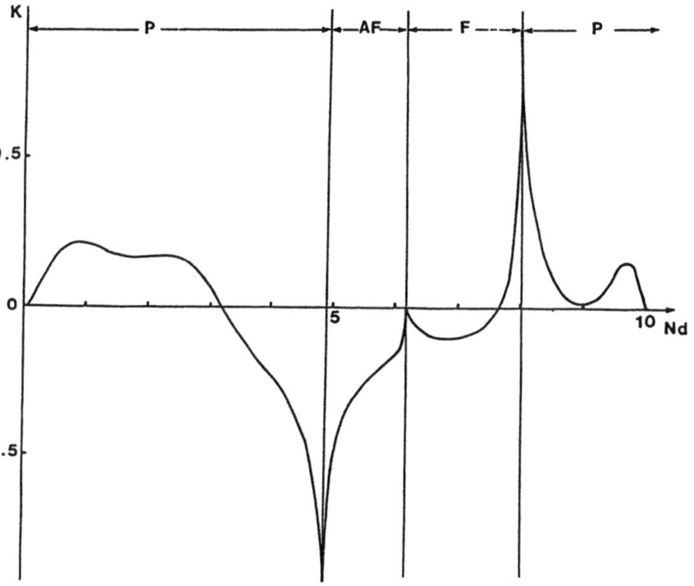

Figure 6. Values of K ($M_i = M_o K^i$) vs. the filling of the d band (calculated with U = 0.6 eV).

a value of U_s larger than the bulk one U as the properties of a surface atom are certainly between those of a bulk atom and those of a free one. Moreover, vanadium is close to chromium, whose U is generally taken equal to 0.6 eV. Vanadium would be a good example for surface live layers. No measurements have yet been done near a (100) V surface, but we can compare these results with the surface magnetic moment deduced from the vanadium hyperfine particle susceptibility [23]. A 2.4 μ_B magnetic moment per surface atom has been deduced from these measurements. Such a value would be obtained near a (100) surface with U_s equal to 1 eV.

For chromium, the large value of the surface paramagnetic density of states at the Fermi level enhances the (100) surface magnetic moment [24] (2.8 μ_B instead of 0.6 in the bulk). Because of the exponential decrease into the bulk of the surface magnetic moment, the net magnetic moment near the surface is not equal to zero as for an antiferromagnetic infinite crystal. Such a surface ferromagnetic layer has been observed near chromium surfaces [25] and its existence suggested from measurements of Cr thin film resistivity [26]. Near a grain boundary there are dangling bonds which may produce effects similar to those at a surface and may induce a magnetic "grain boundary" layer. Electron capture spectroscopy measurements [27] have also revealed a Cr ferromagnetic (100) surface.

Direct evidence of the large Cr(100) surface magnetic moment has recently been obtained using angle-resolved photoelectron spectroscopy [25]. Both majority and minority (100) Cr surface densities of states exhibit surface peaks (labelled 1 and 2 in Figure 7, respectively). These surface peaks correspond to states strongly localized in the surface plane. Their energies are close to the surface intraatomic value of the potential, $V_o \pm V_s$, where V_o is the Coulomb and V_s the exchange potential. So their surface peak separation is just equal to $2V_s$ and proportional to the surface magnetic moment. The magnetic moment vanishes at the Néel temperature so the peak positions must vary with temperature.

Angle-resolved photoelectron spectroscopy measures the sum of the majority and minority spin densities. Moreover, one can estimate the position of the peak 2 to estimate the peak exchange splitting (Figure 8). More recent and accurate density of states calculations have confirmed these results [29]. They even find a larger

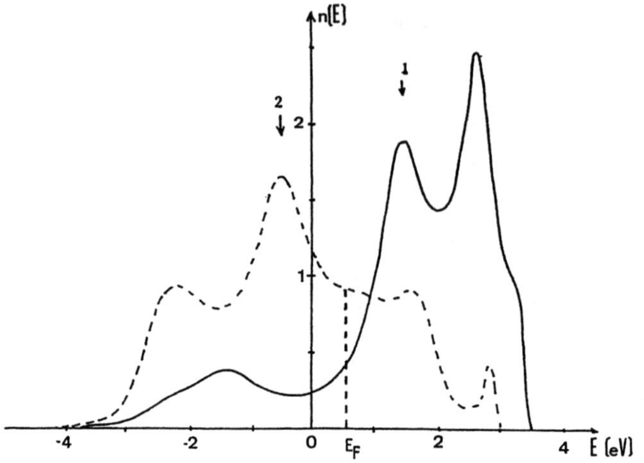

Figure 7. Majority (dotted line) and minority (full line) spin local (100) Cr surface densities of states.

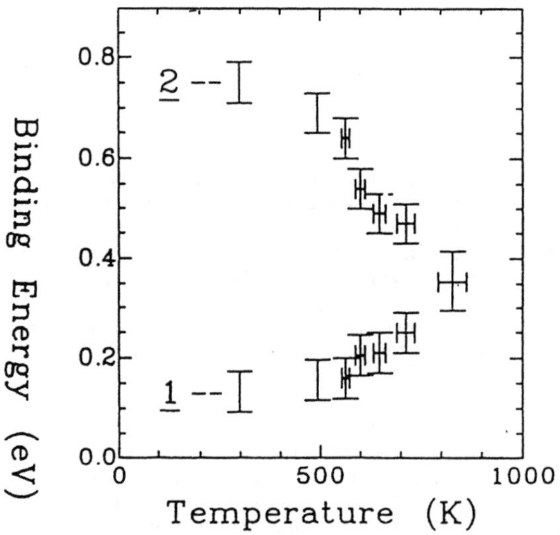

Figure 8. Binding energies of the surface features 1 and 2 plotted vs. temperature [28].

surface magnetic moment (~3.0 μ_B) which decreases as one goes into the bulk with an exponential factor K close to 0.5.

4. Conclusion

We have only developed here a few aspects of surface magnetism. Much progress has been accomplished in the last few years, leading to good agreement between experiment and theory. Some points are not definitively settled. One of them is, for example, the case of a possible silicon (111) surface antiferromagnetism. Surface spin waves have not been observed, although calculations have been done based on itinerant-electron magnetism theory. It is evident that experimental difficulties still remain to be solved. Tools used to study bulk magnetism are either not yet suffiently surface sensitive or they are difficult to set up in the ultra-high vacuum chambers used by surface physicists.

References

1. G. Allan, Surf. Sci. Rep. 1, 121 (1981).
2. J. C. S. Levy, Surf. Sci. Rep. 1, 39 (1981).
3. V. Gradmann, J. Magn. Magn. Mat. 6, 173 (1977).
4. A. J. Freeman, J. Magn. Magn. Mat. 15-18, 1070 (1980).
5. L. N. Liebermann, D. R. Fredkin, and H. B. Shore, Phys. Rev. Lett. 22, 539 (1969); L. N. Liebermann, J. Clinton, D. M. Edwards, and J. Mathon, Phys. Rev. Lett. 25, 232 (1970).
6. G. A. Martin and B. Imelik, Surf. Sci. 42, 157 (1974).
7. G. Bergmann, Phys. Today 32 (8), 25 (1979); Phys. Rev. B 23, 3805 (1981); Phys. Rev. Lett. 41, 264 (1978); Phys. Rep. 107, 1 (1984).
8. J. Tersoff and L. M. Falicov, Phys. Rev. B 26, 6186 (1982).
9. D. Wang, A. J. Freeman, and H. Krakauer, Phys. Rev. B 24, 1126 (1981).
10. G. Allan and P. Lenglart, J. Phys. (Paris) 35, 91 (1974).
11. G. Bergmann and G. Mathon, Solid State Comm. 55, 522 (1985).
12. R. E. Peierls, Ann. Inst. Henri Poincaré, Helv. 5, 177 (1935).
13. L. D. Landau, Phys. Z. Sowjet. 11, 26 (1937).
14. I. S. Jacobs and C. P. Bean, Magnetism, Vol. III, ed. by T. Rado and H. Suhl (Academic Press 1963), p. 301.
15. L. Gunther, Phys. Lett. 25A, 649 (1967); 26A, 216 (1968).
16. S. Alvarado, M. Campagna, and H. Hopster, Phys. Rev. Lett. 48, 51 (1982).
17. W. Eib and S. F. Alvarado, Phys. Rev. Lett. 37, 444 (1976).
18. D. L. Mills and A. A. Maradudin, J. Phys. Chem. Sol. 28, 1855 (1967).
19. G. Allan and M. Lannoo, Surf. Sci. 63, 11 (1977); C. B. Duke and W. K. Ford, Surf. Sci. 111, L685 (1981); R. Del Sole and D. J. Chadi, Phys. Rev. B 24, 7430 (1981); D. Louis, C. Tejedor, and F. Flores, Solid State Comm. 47, 939 (1983).
20. D. Spanjaard, D. L. Mills, and M. T. Beal-Monod, J. Low Temp. Phys. 34, 307 (1979).
21. G. Allan, Electronic Structure and Reactivity of Metal Surfaces, ed. by E. G. Derouane and A. A. Lucas, NATO Advanced Study Institute Ser. (1975), p. 57.
22. G. Allan, Phys. Rev. B 19, 4774 (1979).
23. M. Akoh and A. Tasaki, J. Phys. Soc. Jap. 42, 791 (1977).
24. G. Allan, Surf. Sci. 74, 79 (1978).
25. P. E. Ferguson, J. Appl. Phys. 49, 2203 (1978).
26. H. Hara and M. Sakata, J. Phys. Soc. Jap. 43, 469 (1977).
27. C. Rau and S. Eichner, Phys. Rev. Lett. 47, 939 (1981).
28. L. E. Klebanoff, S. W. Robey, G. Liu, and D. A. Shirley, Phys. Rev. B 30, 1048 (1984).
29. R. H. Victora and L. M. Falicov, Phys. Rev. B 31, 7335 (1985).

Magnetism at the Surface of Transition Metals

K.H. Bennemann

Institute for Theoretical Physics, Freie Universität Berlin, Arnimallee 3, D-1000 Berlin 33, Germany

The spin-fluctuation theory for itinerant electrons is used to calculate the magnitude of the magnetic moments and the temperature dependence of the magnetization at the surface of Fe, Ni, and Cr. We also present results for the effect of chemisorbed and absorbed oxygen on surface magnetism.

1. Introduction

The study of magnetism at the surface of transition metals such as Ni, Fe, and Cr for example, is both of fundamental and technological interest. In particular, the surface might be considered as a special case for testing electronic theories developed recently for magnetism caused by itinerant electrons [1]. It is important to understand whether the Hubbard Hamiltonian, with only intraatomic Coulomb interactions between d electrons, is sufficient to describe fully the change in the magnetic properties resulting from the reduction of the atomic coordination number at the surface. In general we expect that surface magnetism is determined most importantly by the surface shifts of the electronic states, the (mainly) intraatomic s-d charge transfer, change in the s-d hybridization, and the band-narrowing resulting from the reduction in the coordination number. Previous theories [2] studying surface magnetism by using the Heisenberg Hamiltonian and taking the exchange integral J as a phenomenological parameter are difficult to interpret in terms of the above electronic parameters. In the following we use the recently developed spin-fluctuation theory for magnetism caused by itinerant d electrons [1] to calculate the magnetic moments $\mu(T)$ and the magnetization $M_i(T)$ (i refers to the ith atomic plane parallel to the surface) at the surface of Fe(Ni) and Cr. Note, recent calculations performed by Hasegawa for Ni using the electronic spin-fluctuation theory [3] yielded for the case $\mu_s \approx \mu_b$ and spin-averaged neutral atoms $T_{cs} \approx 2T_c$, where T_{cs} refers to the surface Curie temperature. This large enhancement of the Curie temperature T_c results mainly from the band narrowing at the surface. Intuitively (in view of the fact that $T_c \propto zJ$) one expects that the reduction in coordination number z should not cause such a large enhancement of T_c (if $\mu_s \approx \mu_b$). Experimentally one observes for the case of Fe(110) that $M_s(T) > M_b(T)$ at lower temperatures [4], that $M_s(T)$ decreases approximately linearly with T [5], and that $T_{cs} \approx T_c$ [6]; and for the case of Ni(110) one observes [7] $M_s(T) < M_b(T)$, $T_{cs} \approx T_c$ [8], and $\mu_s \approx \mu_b$. Spin-correlations are observed at the surface above T_c up to 2 T_c [9]. Oxygen chemisorbed at the surface of Fe [10] or Ni [11] reduces the surface magnetization. For the case of Cr one observes a ferromagnetic (100)-surface plane, a large enhancement of the magnetic moments, $\mu_s = 2.8\ \mu_b$, and that spin order persists up to $T_c \approx 780$ K [12] ($T_N = 312$ K). Furthermore MEIER et al. [13] observed no surface magnetism if approximately a monolayer of nitrogen or oxygen is chemisorbed at the Cr surface. However, if oxygen penetrates into the surface, one observes again $M_s(T) > 0$ ($M_s \propto c_o$, where c_o denotes the concentration of absorbed oxygen). We attempt to explain these important experimental results in the following by using the electronic spin-fluctuation theory.

2. Theory

A brief outline of a simplified version of the electronic spin-fluctuation theory is presented in the following. It has been shown that this simplified theory works reasonably well for Fe (and Cr), but less so for Ni. For details of the theory see MORAN-LOPEZ et al. [14]. The simplified theory is physically more transparent than more accurate versions of the spin-fluctuation theory [1] and permits in a straightforward way to include spin correlations which may be of particular interest in the study of surface magnetism. The electrons are described by the (tight-binding-like, Hubbard) Hamiltonian

$$H_d = H_s + H_d + H_{sd} \;, \tag{1}$$

where H_s is the usual Hamiltonian for conduction s-electrons, H_{sd} describes the s-d hybridization, and

$$\begin{aligned} H_d = & \sum_{i,\sigma} \varepsilon_{di}\, n_{di\sigma} + \sum_{i,j,\sigma} t_{ij}\, d^{+}_{i\sigma} d_{j\sigma} \\ & + \sum_{i,\sigma} U_i\, n_{i\sigma} n_{i-\sigma} + \text{h.c.} \;+\; \text{(interatomic exchange)} \end{aligned} \tag{2}$$

describes the d electrons. Here, ε_{di} is the d-electron energy in the ith atomic plane, U the intraatomic Coulomb interaction amongst the d electrons, and t_{ij} the matrix element for electronic transitions between atoms at sites i and j. In the following we do not take into account interatomic exchange coupling and assume a 5-fold degenerate d band. According to the decomposition of the surface into atomic planes i parallel to the surface (see Figure 1), one obtains for the free energy of the d electrons

$$F = \sum_i E_i - T \sum_i S_i \;, \tag{3}$$

where E_i is the internal energy contribution from the ith plane and given by

$$E_i = \sum_{\alpha,\sigma} \int d\varepsilon (\varepsilon - \varepsilon_{i\alpha\sigma}) N_{i\alpha\sigma}(\varepsilon) + \Delta E \;. \tag{4}$$

(ΔE describes correlation energy corrections which may be given by the expressions due to FRIEDEL et al. [15]). Here, $\varepsilon_{i\alpha\sigma}$ is the energy of the d state in the ith layer for an atom of kind α, and $N_{i\alpha\sigma}(\varepsilon)$ is the local density of states for d electrons with spin σ in the ith plane at an atom of kind α. We call an atom with magnetic moment parallel to the magnetization of kind A and one with a magnetic moment antiparallel to the magnetization of kind B (Ising-like alloy model), e.g., α = A,B. Then the entropy S_i is given by (the configurational entropy)

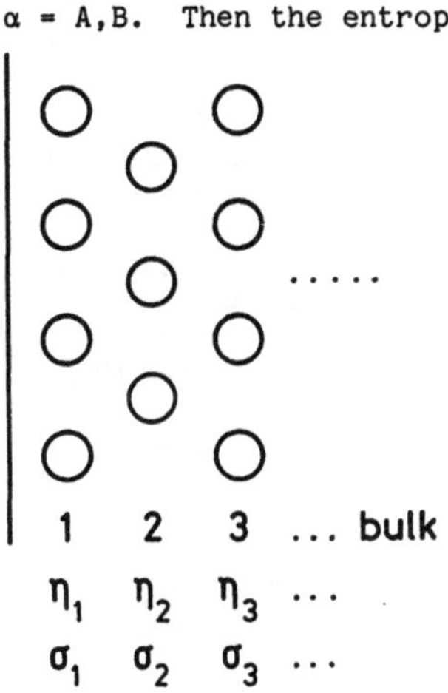

Figure 1. Illustration of decomposition of the transition-metal surface into atomic planes i (i=1,2,3...) parallel to the surface. η_i and σ_i are the long-range and short-range order parameters for the ith plane.

$$S_i = -k_B N_i (P_i^+ \ln P_i^+ + P_i^- \ln P_i^-) , \quad (5)$$

where (N_i = numbers of atoms in i^{th} plane) P_i^+ (P_i^-) is the probability of finding an A (B) atom on site i. If one includes spin pair correlations, this expression for S_i has to be replaced by Kickuchi's expression, such that $P_i \to P_{ij}^{\alpha\beta}$, where $P_{ij}^{\alpha\beta}$ gives the probability to find on site j a β atom if on site i one has a α atom. The (short-range) spin-correlation parameter σ_{ij} is given by

$$\sigma_{ij} = 1 - 2(P_{ij}^{+-} + P_{ij}^{-+}) . \quad (6)$$

The parameter σ_{ij} describes spin-correlations between moments μ_i and μ_j. Note, $P_{ij}^{\alpha\beta}$ or η and σ_{ij} enter F through Kickuchi's expression for S and also through $N_{i\alpha\sigma}(\varepsilon)$. From

$$\frac{\partial F}{\partial \sigma} = 0, \quad \frac{\partial F}{\partial \eta} = 0 , \quad (7)$$

one obtains a coupled set of equations for $\sigma(\eta,T)$ and $\eta(\sigma,T)$. Here, η_i is the layer-dependent long-range order parameter, $\eta_i = P_i^+ - P_i^-$, which is proportional to the magnetization.

The magnetic moment $\mu_i^\alpha(T)$ is determined by (α = A,B)

$$\mu_i^\alpha(T) = \int d\varepsilon f(\varepsilon) \, (N_{i\alpha\uparrow}(\varepsilon) - N_{i\alpha\downarrow}(\varepsilon)) . \quad (8)$$

The quantity $N_{i\alpha\sigma}(\varepsilon)$ follows from the electronic Green's function $G_{i\alpha\sigma}(\varepsilon)$, which is calculated by means of a continued-fraction technique or related approximations [17]. Note that at T = 0 the treatment of the electron-electron interaction is equivalent to the Hartree-Fock approximation.

The above theory is easily extensible to the case of chemisorbed oxygen or nitrogen atoms. Neglecting direct coupling of the chemisorbed atoms, then these influence surface magnetism essentially via the hopping matrix element t_{pd}, describing the hybridization between the p orbitals of the chemisorbed atom and the d states of the substrate transition metal. The resultant change of $G_{i\alpha\sigma}(\varepsilon)$ is determined by adding $H_{pd} = \sum_{i,j} t_{pd} \, d_i^+ p_j$ + h.c. to (1).

To treat antiferromagnets like Cr one introduces two sublattices and then determines the electronic Green's function $G_{i\alpha\sigma}(\varepsilon)$ for the two sublattices. Here, i refers now to the atomic site on one of the two sublattices. Also $P_{ij}^{\alpha\beta}$ or P_i^a includes as usual reference to the particular sublattice.

3. Results

First, we applied the above theory to calculate $\mu_1(T)$ and $M_1(T)$ (i = 1 = surface plane) at the (110)-surface of body-centered-cubic iron. We assume a 5-fold degenerate d band, a number of d electrons n_d = 7.4 , U $\approx$ 1 eV, and a d band width of 4 eV. Spin correlations are neglected. Hence, we put σ = 0. Furthermore, we assume neutral atoms A and B. The results obtained for μ, T_c, and M(T) are shown in Table 1 and Figure 2 [case (a)]. Case (b) refers to calculations by LINKE [17] using for the determination of $G_{i\alpha\sigma}(\varepsilon)$ at the surface a Bethe lattice and determining the free energy F by the functional-integral method used also by HASEGAWA [1]. Note that in Figure 2 we also show for comparison results for M(T) obtained for the bulk, both in the σ = 0 approximation and including spin correlations. The results for $M_b(T)$ and $\sigma \neq 0$ are in excellent agreement with experimental results, in contrast to previous calculations by HUBBARD and also by HASEGAWA [1]. From the calculations performed for the bulk we conclude that also at the surface short-range spin correlations occur over a distance ℓ = 3 ÷ 5 Å at temperatures of the order of T_c and above T_c. Spin correlations tend to couple better the surface plane to the bulk. Therefore we expect that (T_{cs}/T_c) will decrease if we include σ_{ij} in the calculations.

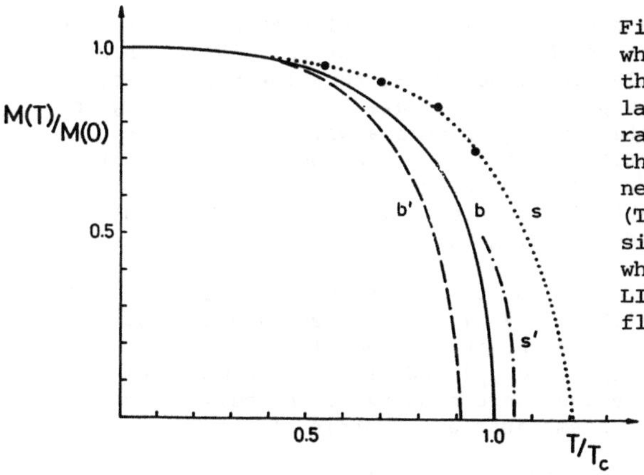

Figure 2. Magnetization M(T) for Fe, where s refers to the (100) surface, b to the bulk without short-range spin correlations, and b' to the bulk with short-range spin correlations included. T_c is the bulk Curie temperature calculated neglecting short-range spin correlations ($T_c \approx 2000$ K). Curve s results from the simple alloy theory for spin fluctuations while curve s' gives results obtained by LINKE using functional integral spin fluctuation theory.

Table 1. Results for the magnetic moment μ_1 and the Curie temperature T_{cs}
($T_{cb} \approx 2000$ K). Case (b) refers to calculations performed by LINKE not using the configurational entropy expression used in case (a) at the (110)-surface of body-centered-cubic iron.

	a	b
μ_s/μ_b	~1.1	1.06
T_{cs}/T_c	~1.2	1.05

Summarizing, we note that $\mu_s \simeq \mu_b$ and also $T_{cs} \geq T_{cb}$. Also, $M_s > M_b$; however, in contrast to experiment [5] we do not obtain $M_s(300) \simeq 1.3\ M_b(300)$ nor a linear decrease with T. Presumably, we have to calculate a better D.O.S. and to include spin correlations and maybe interatomic exchange interactions J_{ij} to obtain a flattened M(T) curve. Corresponding calculations are in progress. The physical reason for $\mu_s > \mu_b$ and $T_{cs} > T_c$ seems to be the narrowing of the majority and minority spin bands, which cause an increase of $n_{d\uparrow}$ and a decrease of $n_{d\downarrow}$. For $n_{d\uparrow}$, for example, the s-d hybridization plays a role by affecting the tail in $N_{d\uparrow}(\varepsilon)$ above the Fermi level ε_F.

Preliminary results for Ni yield also $\mu_s \leq \mu_b$ and $T_{cs} \simeq T_{cb}$. Both for Ni and Fe these results depend sensitively on the conditions for electronic charge transfer. We also calculated T_{cs} from $T_c = z\mu^2 J$ by replacing the surface coordination ($z = 4$), the calculated values of μ_s and by determining the exchange integral J the same way Hubbard did for Ni, $J = (E_- - E_+)/zM(\mu^+ + |\mu^-|)$. Again, we obtain $T_{cs} \approx T_{cb}$. Note that if one assumes surface level shifts to lower binding, analogous to the surface core level shifts, then one expects already the above obtained results for Fe, and for Ni possibly $\mu_s > \mu_b$.

The calculations performed for a monolayer of O on Fe(110), assuming O^- and mainly p-d hybridization near ε_F yielded approximately the results shown in Table

Table 2. Reduction of μ_s(Fe) due to chemisorbed O^- ($\mu_{ox} = 1\ \mu_B$). Results (a) are obtained for $t_{pd} = 3$ eV and (b) for $t_{pd} = 1$-2 eV.

	a	b
μ_s	0.8 μ_b	0

2. We use d $\approx$ 1 Å for the oxygen-iron distance. The reduction of μ_s results from the hybridization between the oxygen p- and Fe 3d states near ε_F. For O^- this will essentially pull down some minority spin states below ε_F, thus lowering μ (Fe). However, more detailed calculations (in progress) indicate that both minority and majority spin states near ε_F are affected by t_{pd} in such a way that μ_s is lowered [18]. If oxygen is adsorbed on interstitial sites near the surface, we expect again strong effects on μ and M(T). For absorbed oxygen, strong local lattice disorder occurs, since $O \rightarrow O^{2-}$ presumably if on interstitial sites. This lattice distortion will play a strong role in determining μ_s.

Finally, we applied the theory outlined in Section 2 to the surface of Cr. We assume a 5-fold degenerate d band of width W = 4 eV, U = 0.6 eV, and n_d = 5.4. Then, as in previous calculations [12,19] we obtain for the clean Cr(100) non-reconstructed surface $\mu_s \approx 2.8\ \mu_B$ ($\mu_b = 0.6\ \mu_B$) and also an enhanced moment μ_2. The spin-order configuration is the one corresponding to the bulk. The ferromagnetic spin order at the surface seems just a geometric effect.

Assuming a monolayer of oxygen on four-fold chemisorption sites, a charge transfer such that O^- is present, and no direct electron hopping between the oxygens, then depending on the p-d hybridization described by the matrix-element t_{pd}, one obtains a significant reduction of μ_s. Taking for simplicity $\mu_2 = \mu_2(\theta = 0)$, then approximately [18] $\mu_s = 0.2 \div 0.3\ \mu_B$ for t = 1 ÷ 2 eV. This demagnetization gets stronger if the distance d between the surface and the chemisorbed atom decreases. This may be the case for N which is observed to sit closer to the surface and to penetrate into the surface.

If oxygen is absorbed on interstitial sites near the surface, then the charge transfer increases, $O \rightarrow O^{2-}$, and consequently strong local lattice distortion occurs. The latter may be estimated from the atomic radii of O^{2-} and Fe. Preliminary results are $\mu \approx 1.2\ \mu_B = 2\ \mu_b$ for a Cr atom next to an oxygen atom on an octahedral intersitial site, if lattice distortion is included (otherwise $\mu \simeq 1.2\ \mu_b$). This gives per oxygen atom a moment enhancement of 2.4 μ_B as compared to 4 μ_B estimated from experiment [13]. Final results will be published elsewhere [18].

If we treat now the surface as a magentic alloy (see Figure 3) with a concentration c_o of impurity moments μ_i representing the total moment of Cr atoms which sit around an interstitial oxygen atom and other moments μ due to the Cr atoms which are not affected by oxygen, then a usual molecular field type calculation with a Heisenberg Hamiltonian $H = -\Sigma J S_i \cdot S_j$ yields [18] ferromagnetic order of the cluster moments μ_i and a Curie-temperature $T_c \approx 530$ K for c_o = 3% (and if μ_i = 2.4 μ_B is assumed) which compares well with experiment [13].

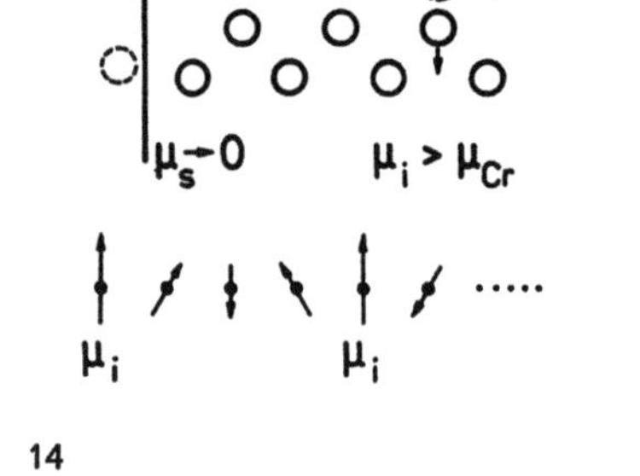

Figure 3. Illustration of the Cr surface with chemisorbed and absorbed oxygen. μ_i denotes the total magnetic moment of displaced Cr atoms, nearest neighbors to oxygen. The lower part of the figure illustrates the alloy consisting of spins μ_i and μ_{Cr} used to calculate T_c.

In summary, some results for μ, M(T), and T_c obtained by means of an electronic theory are presented. Further studies are needed to understand whether a Hubbard type Hamiltonian treated within the Hartree-Fock (or somewhat better) approximation is sufficient to describe surface effects on the magnetism of transition metals.

I am greatly indebted to R. Linke, A. Aligia, J. Dorantes-Dávila and G. Pastor for some of the results and for useful discussions.

References

1. Electron Correlation and Magnetism in Narrow Band Systems, ed. by T. Moriya (Springer, Berlin, Heidelberg, New York, 1981); J. Hubbard, Phys. Rev. B 19, 2626 (1979); Phys. Rev. B 20, 4584 (1979); Phys. Rev. B 23, 5974 (1981); H. Hasegawa, J. Phys. Soc. Jap. 46, 1504 (1979).
2. K. Binder and P. C. Hohenberg, Phys. Rev. B 9, 2194 (1974); S. Stelzer and N. Maglin, Phys. Rev. B 26, 404 (1982).
3. H. Hasegawa, private communication and to be published.
4. E. Tamura and R. Feder, Solid State Commun. 44, 1101 (1982).
5. S. Alvarado, this volume, private communication and to be published.
6. J. Kirschner, M. Glöbe, V. Dose, and H. Scheidt, Phys. Rev. Lett. 53, 612 (1984).
7. S. Alvarado, M. Campagna, and H. Hopster, Phys. Rev. Lett. 48, 51 (1982); R. Feder, S. F. Alvarado, E. Tamura, and E. Kisker, Surf. Sci. 127, 83 (1983).
8. R. J. Celotta, D. T. Pierce, G.-C. Wang, S. D. Bader, and G. P. Felcher, Phys. Rev. Lett. 43, 728 (1979).
9. C. Rau and S. Eichner, Phys. Rev. Lett. 47, 939 (1981).
10. R. Allenspach, M. Taborelly, and M. Landolt, Phys. Rev. Lett. 55, 2599 (1985).
11. D.T. Pierce, private communication and to be published.
12. L. E. Klebanoff, S. W. Robey, G. Liu, and D. A. Shirley, Phys. Rev. B 30, 1048 (1984); L. E. Klebanoff, R. H. Victora, L. M. Falicov, and D. A. Shirley, Phys. Rev. B 32, 1997 (1985).; D. R. Grempel, Phys. Rev. B 24, 3928 (1981).
13. F. Meier, D. Pescia, and T. Schriber, Phys. Rev. Lett. 48, 645 (1982).
14. J. L. Morán-López, M. Avignon, and K. H. Bennemann, Phys. Rev. B 23, 5978 (1981).
15. J. Friedel, in The Physics of Metals, ed. by J. M. Ziman (Cambridge, 1969), chapter 8.
16. R. Kickuchi, Phys. Rev. 81, 988 (1951).
17. R. Linke, Thesis, FU Berlin, 1985; J. Dorantes-Dávila and G. Pastor, private communication and to be published.
18. A. Aligia and K. H. Bennemann, private communication and to be published.
19. G. Allan, Surf. Sci. 74, 79 (1978).

Magnetism at Surfaces and Interfaces

A.J. Freeman and C.L. Fu

Department of Physics and Astronomy, Northwestern University,
Evanston, IL 60201, USA

1. Introduction.

As is clear from this International Workshop, the study of magnetism at surfaces and interfaces has added new excitement to this largely unexplored frontier field. Much of this theoretical and experimental effort has been addressed to transition-metal materials and represents part of a general increase in interest in the study and understanding of these materials on the microstructure scale. The unique chemical and physical properties of transition metals play an especially important role in determining the observable phenomena in such reduced symmetry systems as small particles (or clusters of atoms), surfaces, interfaces, and modulated structures. A driving force in this exciting area is that new insights into their electronic structure and magnetism are expected because (i) the existence and role of surface and interface states and (ii) the reduced coordination and symmetry lead to important property differences with respect to bulk systems. In particular, the possibility of magnetically dead layers [1,2] at the surface of ferromagnetic Fe and Ni invoked considerable discussion and stimulated the development of theoretical methods for describing surface electronic structures.

Compared to the number of bulk calculations, there have been relatively few such studies for surface and other reduced symmetry systems. This paper concerns itself with giving an indication of the present state of ab initio self-consistent local spin density determinations of the electronic structure and magnetism at surface and interfaces. Of necessity -- because of spâce limitations -- we will restrict ourselves essentially to the work in which we have been involved. (Abundant references to other work are given in the papers cited here. In this way, the reader will have what we hope is a coherent overview of developments in this field.)

2. Approach and Methodology.

A. Thin-Slab Approximation.

Our focus in this paper is the treatment of extended surfaces with or without ordered overlayers of absorbed atoms. Probably the most successful structural model uses a thin-slab to simulate both surface and bulk effects on an equal footing. Calculations in this model are performed for a finite thickness slab which is infinitely periodic in the plane parallel to the surface. Typically, slabs 5-13 atomic-layers thick are used. The features of this model have been discussed at length by APPELBAUM and HAMANN [3]. To ensure accurate results, the slab should be thick enough so that the electronic structure in the interior of the slab resembles closely the expected bulk structure. Thicker slabs also reduce size-effect energy splittings between surface states which are localized on the upper and lower surfaces. For metals with well localized surface states (such as transition metals), these requirements are satisfied for slabs as thin as five layers. If it is necessary to identify more extended surface states, thicker slabs must be used. The desirability of thicker slabs for the above reasons must be balanced, however, by the practical consideration that the magnitude of the calculation (measured by the amount of computer memory and computer time required) increases rapidly with increasing slab thickness (roughly between N^2 and N^3, where N is the number of layers in the slab).

B. FLAPW Method for Thin Films.

It is now recognized that density functional theory [4] provides an elegant and powerful framework to describe the electronic structure of condensed systems (e.g. bulk crystals, surfaces, interfaces). In its local approximation density functional theory leads to Schrödinger-like one-particle equations (Kohn-Sham equations) containing an effective potential energy operator which is determined by the self-consistent charge distribution. Thus, the local density functional one-particle equations have to be solved iteratively.

One of the most precise and powerful schemes to solve the local (spin) density (LSD) one-particle equations for the film geometry is the all-electron, full-potential linearized-augmented-plane-wave (FLAPW) method [5]. The basic idea in this variational method is the partition of real space into three different regions, namely spheres around the nuclei, vacuum regions on both sides of the film, and the remaining interstitial region. In each of these regions the "natural" form of the variational basis functions is adopted, i.e., plane waves in the interstitial region, a product of radial functions and spherical harmonics inside the spheres, and in the vacuum a product of functions which depend only on the coordinate normal to the film and two-dimensional plane waves. Each of these basis functions is continuous in value and derivative across the various boundaries. This is possible because inside the spheres (and analogously in the vacuum) two radial functions for each ℓ-value are used, namely the solution of the radial Schrödinger equation for the correct potential and its energy derivative.

In the FLAPW method no shape approximations are made to the charge density and the potential. Both the charge density and the effective one-electron potential are represented by the same analytical expansions described above, i.e., a Fourier representation in the interstitial region, an expansion in spherical harmonics inside the spheres, and in the vacuum two-dimensional Fourier series in a set of planes parallel to the surfaces. The generality of the potential requires a method to solve Poisson's equation for a density and potential without shape approximations. This is achieved by a new scheme which goes beyond the Ewald method [6]. The key idea in this new scheme is the observation that the potential outside a sphere does not depend on the actual charge density inside the sphere but only on its multipole moments. Now, Poisson's equation is solved straightforwardly when the charge density is given in a Fourier representation. Because of the sharp structure of the charge density in the core region (including the nuclear charge), a Fourier expansion of the total density would be extremely slowly convergent. However, since the potential outside the sphere depends on the charge inside only through the multipole moments, the true charge density can be replaced by a smooth density which has a rapidly converging Fourier series and the same multipole moments as the true density. With this replacement of the density inside the spheres we have a Fourier expansion of a charge density which gives the correct potential outside and also on the sphere boundaries. To find the potential inside the sphere we are faced in a final step with a standard boundary-value problem of classical electrostatics which can be solved from the original charge densities inside the spheres and the potential on the sphere boundaries by a Green's function method.

Thus, the FLAPW method allows a fully self-consistent solution of the LSD one-particle equations for the film geometry and yields charge densities and spin densities close to the LSD limit. Beside the total charge density, the key quantity in density-functional theory is the total energy corresponding to the ground-state charge density. Recently, we have presented a new scheme [7] to calculate highly precise and stable all-electron density functional total energies and have applied it within the FLAPW method. The capability of total energy calculations for various geometrical arrangements provides us with a powerful theoretical tool to study the energetics and, at least in principle, the dynamics of surfaces and overlayers.

3. Magnetism of Transition-Metal Surfaces.

The FLAPW approach has now been applied to the study of the magnetism of a number of transition-metal surfaces. Here we are able to present only a small sample of these results.

A. Fe(001) and Fe(110).

Consider first the case of Fe(001). In Figure 1 we show the spin density of the Fe(001) surface for the (110) plane perpendicular to the surface. The greatest part is dominated by positive spin densities with only small pockets of negative spin densities between the atoms. In the bulk-like center the shape of the spin density shows important nonspherical components of t_{2g} symmetry. Surprisingly, this bulk-like shape is found even for the iron atoms just one layer below the surface, and only the surface atoms exhibit a different shape of the spin density which is markedly more spherical. The polarization on the vacuum side of the surface atoms is found to be positive.

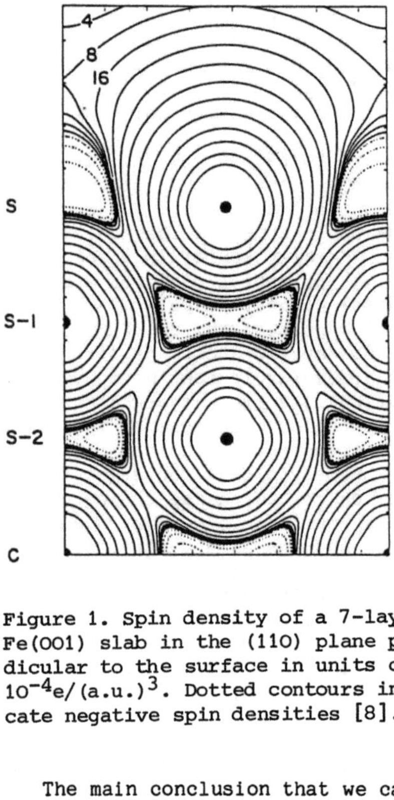

Figure 1. Spin density of a 7-layer Fe(001) slab in the (110) plane perpendicular to the surface in units of $10^{-4}e/(a.u.)^3$. Dotted contours indicate negative spin densities [8].

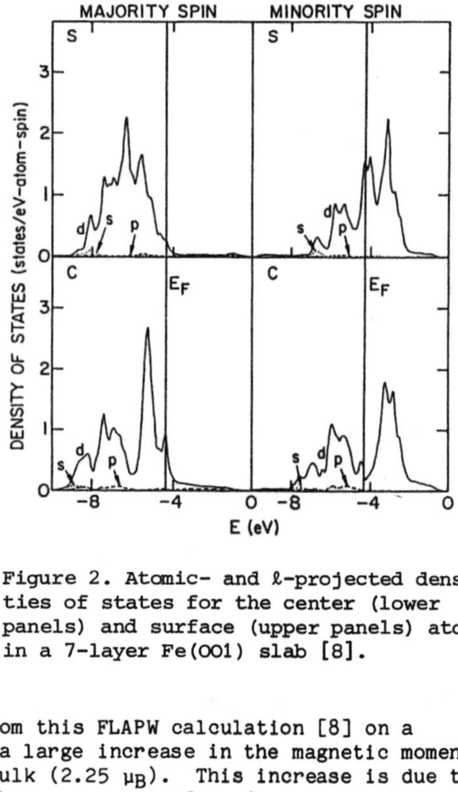

Figure 2. Atomic- and ℓ-projected densities of states for the center (lower panels) and surface (upper panels) atoms in a 7-layer Fe(001) slab [8].

The main conclusion that we can draw from this FLAPW calculation [8] on a seven-layer Fe(001) film is that there is a large increase in the magnetic moment at the surface (2.98 μ_B) compared to the bulk (2.25 μ_B). This increase is due to the surface-induced d-band narrowing and the occurrence of surface states below the Fermi energy (c.f. Figure 2). (The agreement between experiment [9,10] and theory for the surface states is quite good.) Not only are "dead" layers ruled out, but a simple picture of the magnetism of the clean surfaces emerges. In this picture, the magnetization is a monotonic function of the dimensionality, with the bulk at one end and the free atom at the other. This simple picture is supported by self-consistent calculations (and experiments) on bulk, surfaces, monolayers, linear chains [11], and atoms. Because of the large moment in Fe(001), it is not surprising that the magnetism also has an effect on the electronic properties. One example is the dependence of the work function Φ (a mainly electrostatic property) on the allowed magnetic order. A paramagnetic (i.e. non-spin-polarized) film yields Φ = 4.86 eV, whereas the spin-polarized result is 4.29 eV. (The experimental value is 4.4 eV) [9,10]. This result is easily understood: ferromagnetic ordering fills the more compact antibonding type states at the expense of bonding states, resulting in a dipole barrier (relative to the paramagnetic surface) such as to reduce Φ. Obviously this mechanism is important only if there is a large moment, as in the present case.

The surface magnetism of Fe(110) has been well-studied using various experimental techniques [12,13]. In contrast to the open character of the (001) surface, the (110) surface has a higher surface packing density, i.e., six nearest neighbors for each surface atom. Nevertheless, results based on SPLEED and dynamical theory [12] predicted that (i) the surface magnetism of Fe(110) is enhanced by 30% and (ii) a layer-dependent Friedel-type oscillation in the magnetization. Recently, magnetic hyperfine fields at nuclei (H_{hf}) near the surface have been measured by means of Mössbauer spectroscopy [13].

We have carried out an FLAPW local-spin-density functional study of the magnetic and electronic structures of Fe(110) [14]. In order to eliminate possible thin film size effects, the surface was represented by a nine-layer single slab (18 atoms/unit cell). This FLAPW investigation shows: (1) the magnetic moments within the MT sphere of each ferromagnetic layer are found to be (from the central layer to the surface) 2.22, 2.25, 2.28, 2.37, and 2.64 μ_B, i.e., the surface magnetic moment is enhanced by 20%; (2) no Friedel-type oscillation of the spin density is found upon going from the surface into the bulk region; and (3) the surface-induced d-band narrowing is less pronounced compared to that of the (001) surface, and the exchange splitting is found to be ~2.5 eV. It is interesting to note that the surface magnetic moment (2.64 μ_B) lies between those of the bulk and (001) surface, i.e., it scales with the coordination number of surface atoms. Furthermore, we obtained 5.10 eV for the work function of the Fe(110) surface (~0.8 eV larger than that of the Fe(001) surface), which also reflects the more close-packed character of the (110) surface.

Since the valence s-electrons at the (110) surface become less atomic-like, we would expect a decrease of the direct (positive) polarization of the valence contribution to H_{hf}, resulting in an increase in magnitude of H_{hf} (dominated by the negative contribution of core electrons) at the surface compared to the (001) surface [8]. The calculated H_{hf} for (110) surface atoms, 31.0 ± 0.5 T, is found to agree well with Mössbauer measurements [13] (30.88 ± 0.4 T).

B. Ni(001).

Since the report of magnetically "dead" layers on a Ni(001) surface [2], this system has attracted a host of experimental and theoretical investigations, partly with contradicting conclusions. It is now established, however, that the clean Ni(001) surface is not magnetically "dead". This result is based on experiments using spin-polarized field emission [15], spin-polarized photoemission [16], electron capture spectroscopy [17], and polarized electron diffraction [18] and on theoretical calculations of the surface magnetization [19-23]. It seems that the early experimental results showing magnetically "dead" layers have been clouded by difficulties in the sample preparation and characterization, partly due to interface problems.

Recent highly accurate FLAPW studies on a seven-layer Ni(001) slab [23] shed new light onto the magnetism of the Ni(001) surface: (i) the magnetic moments in atomic-nearest neighbor volumes are found to be (from the central layer to the surface) 0.56, 0.59, 0.60, 0.68 μ_B, i.e., there is an enhancement of the magnetic moment of ~20% of the surface compared to bulk; (ii) no Friedel oscillation of the moment going from the surface to the center of the seven-layer film; and (iii) the majority surface state $\bar{M}_3$ lies 0.14 eV below the Fermi energy. i.e.. there is no majority-spin d-hole at $\bar{M}$. The theoretical result for the work function, 5.37 eV, agrees well with the experimental value of 5.22 eV.

It is remarkable that for both Fe and Ni the reduced number of nearest neighbors on the surface leads to an enhancement of the magnetic moment. This trend becomes even more obvious and a simple picture emerges when we include linear chains and free atoms in the consideration: going from bulk to a (001) surface, then to a linear chain [11], and finally to the free atom, the moments are 0.56, 0.68, 1.1, and 2.0 for Ni and 2.27, 2.98, 3.3, and 4.0 for Fe. Thus, as the dimensionality is decreased, the magnetic moments approach the values of the free atoms.

C. Cr(001).

As is apparent also from a number of other papers at this workshop, intense interest exists in the Cr(001) surface. This interest arises because of: (1) the possibility of induced ferromagnetism in its surface layer while retaining antiferromagnetic behavior in the bulk [24-35]; (2) interest in the intriguing surface electronic structure of the (001) surface for the Group VI B transition metals in general and possible phenomena associated with the high density-of-states which arises from localized surface-state bands near the Fermi level in particular. For W(001) [36] and Mo(001) the surface state coupling plays a dominant role in the structural phase transition to the c(2x2) structure, whereas for Cr(001) the relatively narrow 3d surface-state band favors a large enhancement of the local spin susceptibility (i.e., a Stoner instability). For these reasons, one might expect ferromagnetic ordering in the surface layer of Cr(001).

The existence of a ferromagnetic surface layer in antiferromagnetic chromium was first reported with the observation of surface magnetoplasma waves via an attenuated total reflection method [24] on polycrystalline chromium films. It has also been reported [26] that in small particles of chromium magnetic ordering persists well above the Neél temperature at 310 K. Recently, an electron capture spectroscopy study by RAU and EICHNER [27] further confirmed the existence of ferromagnetic order below 365 K, but with a c(2x2) reconstructed surface. MEIER *et al.* [28] found no indication of ferromagnetism in spin-polarized photoemission studies from a chromium surface which, as they noted, included coverage by a nitrogen monolayer on top of the surface.

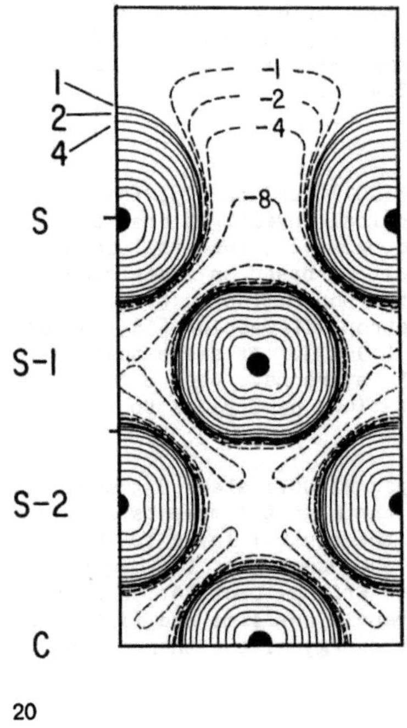

Figure 3. Spin density of a 7-layer Ni(001) slab in the (100) plane perpendicular to the surface in units of $10^{-4}e/(a.u.)^3$ [20].

It was later confirmed by LEED analysis [37] that the c(2x2)-Cr(001) structure is related to surface impurities such as C, N, O, or S, and that clean surface exhibits a p(1x1) LEED pattern at room temperature. Very recently, a surface magnetic phase transition was observed on Cr(001) using angle-resolved photoelectron spectroscopy (ARPES) by KLEBANOFF et al. [30]; from the temperature dependence of the splitting of the surface resonance state, the transition temperature was found to be 780 K, and the surface magnetic moment was estimated to be 2.4 μ_B. These findings are supported by the studies of ZAJAC et al. [38] on the epitaxy and electronic structure of Cr overlayers deposited on single crystal Au(001).

Recent theoretical studies [33] on a seven-layer Cr(001) slab using the FLAPW method confirms a large enhancement of the magnetization at the surface. The calculated spin density for the seven-layer Cr(001) film exhibits antiferromagnetic coupling between adjacent layers in the (001) direction. The induced ferromagnetic state has layer-projected magnetic moments within each MT sphere from the center to the surface layer given as -0.89, 0.89, -1.29, and 2.49 μ_B. This surface-induced magnetic moment of 2.49 μ_B is comparable with (but smaller than) earlier theoretical results obtained from the tight-binding method (2.8 μ_B) by ALLAN [25], from a DVM-LCAO calculation (2.56 μ_B) by HIRASHITA et al. [32], and the value (3.0 μ_B) obtained by VICTORA and FALICOV [34] using a single site Hartree-Fock tight-binding approach using parameters determined from bulk Cr. In addition, the calculated spin polarization shows large penetration of the enhanced surface magnetization into the bulk. (For bulk Cr, the measured maximum magnetic moment is 0.59 μ_B.) As will be discussed later, the calculated FLAPW surface moment is smaller than the moment (3.0 μ_B) found for Au/Cr/Au(001) sandwiches.

Similar to the case of Fe(001) [8], the surface magnetization reduces the work function by 0.35 eV from that of the paramagnetic state (from 4.40 eV to 4.05 eV). While this theoretical value is 0.4 eV lower than the experimental value obtained by WILSON and MILLER [39] (4.46 eV), it is very close to the photoemission result of MEIER et al. [28] (4.10 ± 0.05 eV). Although both experiments claimed to stand for the work function of the clean Cr(001) surface, the agreement with the latter may be fortuitous owing to the presence of one monolayer of nitrogen and zero spin-polarization on their p(1x1) surface. On the other hand, it should be noted that the often quoted experimental value (4.46 eV) [39] refers to a surface contaminated by a submonolayer of impurities (O, S, and N) with a c(2x2) reconstructed structure. Thus, these experimental values cannot be regarded as the work function of a clean Cr(001) surface.

An examination of the DOS at E_F (3 states/eV-atom) suggests ferromagnetic ordering on the surface, since a Stoner instability can be easily achieved even with an exchange integral assumed equal to the bulk value, 0.67 eV. [As found elsewhere for V(001), this sharp surface LDOS peak which is characteristic of the occurrence of surface magnetism in the 3d transition metals is located 0.3 eV above E_F in V(001) [40]. As a result, and as has been confirmed by total energy calculations [40], the paramagnetic state is more stable than the ferromagnetic state for V(001).] The ℓ-decomposed partial LDOS in the surface layer MT sphere for the ferromagnetic state, reveals a large spin-density imbalance and gives rise to an enhancement of the local atomic moment to 2.49 μ_B per surface atom. In agreement with earlier estimates [25,30], our calculated ferromagnetic surface exchange splitting is found to be 1.8 eV.

4. Giant 2D Overlayer and Interface Magnetism.

We have presented some recent results of theoretical FLAPW local density studies of the surface magnetism of transition metals. Since the localized surface states play an important role in the enhancement of surface magnetism for 3d transition metals, much attention has been focussed on the possibility of inducing two-dimensional magnetism in a controlled way (such as in overlayers, sandwiches, and modulated structures). We have, in fact, extended the surface studies reported here to a variety of such overlayer and sandwich systems [41,42]. These extensive

FLAPW studies predicted giant 2D magnetism in a number of transition metal-noble metal systems. Only a brief indication of these results can be given here. For example, when a monolayer of Cr(001) is deposited onto Au(001), the magnetic moment of the Cr overlayer decreases by only a small amount from the free monolayer value (4.12 μ_B) to 3.70 μ_B. This extremely large moment -- the largest value reported for a transition metal other than for Mn -- is 50% greater than that of the surface layer in Cr(001) predicted theoretically (2.49 μ_B) [33] or derived experimentally (2.4 μ_B) [30]. Also surprising is the finding that this enormously enhanced moment is only moderately reduced (to 3.10 μ_B) when the Cr overlayer is itself covered by an Au layer. A very similar result was obtained by Oguchi and Freeman [43], 2.95 μ_B per Cr atom, when the (1x1) Cr/Au coherent modulated structure (superlattice) was studied using the LMTO approach. Since Cr is a notorious getter, the retention of this enhanced 2D magnetization in either the single sandwich or superlattice structures might make its observation much easier. Further, additional theoretical studies show that the Cr magnetic moment is hardly affected by the increase of Au thickness; more layers of Au tend to reduce the antiferromagnetic coupling between the ferromagnetic monolayer sheets in the superlattice geometry [43]. Thus, there is the added degree of freedom of variable wavelength of modulation -- and magnetic coupling -- which may be important for technical applications including magnetic recording. Significantly, there is also a sizeable magnetic moment induced onto neighboring Au atoms (0.10-0.14 μ_B) indicating a large magnetic perturbation of the Au induced by the magnetic Cr atoms. As in the case of the Cr atoms, the polarization of the Au atoms is mainly from the d-like electrons; the contribution from the s- and p-like electrons is only 1-2%. Only one layer away from the interface, the electronic structure of Au is essentially unperturbed and the bulk structure and properties of Au are retained.

Similar investigations [41] have also been performed for monolayers of Fe and V adsorbed on Ag(001). For Fe there is a close matching of the lattice constants but a 5% mismatch for V. As in the case of Cr/Au(001), large magnetic moments, 2.96 μ_B (Fe) and 2.0 μ_B (V), are found for the adsorbate monolayers. The magnetic moment of the Fe overlayer on Ag(001) is remarkably close to the theoretical magnetic moment [8] of the surface layer of an Fe(001) film (2.98 μ_B) -- again indicating the lack of interaction with the substrate.

The result for V/Ag(001) is much more surprising since, like the bulk, the surface layer of V(001) is not magnetic [40]. We thus have the remarkable prediction that an overlayer of V on Ag(001) is magnetically ordered with a sizeable magnetic moment (2.0 μ_B) which is almost as large as the moment of Fe in bulk Fe. [If confirmed, this will be the first solid material for which elemental vanadium demonstrates magnetic ordering.] The origin of magnetic ordering is not negative pressure since, in fact, the matching to the Ag(001) substrate results from a reduced lattice constant for V by 5%.

For both the Fe and V overlayers, it is found -- despite some hybridization between the d-bands of the adsorbate and substrate -- that the adsorbate localized surface state bands retain their quasi-2D behavior for these systems. This behavior is further demonstrated for a monolayer of Fe sandwiched by Ag(001) -- the magnetic moment of Fe is only slightly decreased from that of the overlayer of Fe/Ag(001) to 2.80 μ_B. [Similarly, a moment of 2.92 μ_B on the Fe site is found for an Au/Fe/Au(001) sandwich.] However, when a monolayer of V(001) is sandwiched by Ag(001), the vanadium layer becomes paramagnetic. Apparently, the hybridization with Ag further reduces the DOS of V near E_F and results in a Stoner factor which is just less than one (the exchange correlation parameter for V is 0.4 eV [40]).

The sensitivity of the magnetic ordering of V to its metallic environment is further illustrated by calculations with two layers of V on Ag(001) [41]; in this case the interface layer atoms remain essentially in the paramagnetic state (zero moment) and the surface layer has a moment of 1.15 μ_B which is substantially reduced from that of the single overlayer value (2.0 μ_B). This calculation leads to the result that an Ag substrate is more amenable for the magnetism of the V monolayer than is another V layer.

In order to further illustrate that the noble-metal substrate is more amenable for the magnetism of a 3d metal monolayer sheet than is another magnetic material, we have investigated the electronic and magnetic properties of Cr monolayer on Fe(001) [42]. We found that the Cr and Fe layers tend to couple antiferromagnetically in the <100> direction; further, although there is an enhancement of the Cr moment at the Cr/Fe interface [to 3.10 μ_B accompanied by a decrease of the Fe moment (-1.98 μ_B)] from that of the clean Cr(001) surface, it is 0.6 μ_B less than the magnetic moment of Cr on Au(100). We have also carried out calculations to examine the effect of a nonmagnetic simple-metal substrate on the spin polarization of a Cr monolayer [i.e., Cr/Al(100)]. Although a larger d-sp hybridization is realized in this case, no evidence of a magnetic proximity effect is found; the magnetic moment of the Cr monolayer is largely retained (3.10 μ_B). Again, these analyses indicate that the isolation of the interface states of the 3d transition metal is a key factor in allowing the formation of "giant" magnetic moments.

Finally, we must emphasize that these predictions of strongly enhanced 2D magnetism, thus far on only a fraction of various combinations possible, are part of a growing literature which indicates the range and wealth of magnetic phenomena available for experimental investigation by a variety of techniques. It also indicates the new predictive role of all-electron local spin density theory.

Acknowledgments

We are grateful to our colleagues and collaborators who participated in the work reported here. The research was supported by the U.S. National Science Foundation and the Office of Naval Research.

References

1. L. N. Lieberman, D. R. Fredkin, and H. B. Shore, Phys. Rev. 22, 539 (1969).
2. L. N. Lieberman, J. Clinton, P. M. Edwards, and J. Mathon, Phys. Rev. Lett. 25, 232 (1970).
3. J. A. Appelbaum and D. R. Hamman, Rev. Mod. Phys. 48, 478 (1976).
4. P. Hohenberg and W. Kohn, Phys. Rev. 136, B864 (1964); W. Kohn and L. J. Sham, Phys. Rev. 140, A1133 (1965).
5. E. Wimmer, H. Krakauer, M. Weinert, and A. J. Freeman, Phys. Rev. B 24, 864 (1981), and references therein.
6. M. Weinert, J. Math. Phys. 22, 2433 (1981).
7. M. Weinert, E. Wimmer, and A. J. Freeman, Phys. Rev. B 26, 4571 (1982).
8. S. Ohnishi, A. J. Freeman, and M. Weinert, Phys. Rev. B 28, 6741 (1983).
9. A. M. Turner, Y. J. Cheng, and J. L. Erskine, Phys. Rev. Lett. 48, 348 (1982).
10. A. M. Turner and J. L. Erskine, Phys. Rev. B 28, 5628 (1983).
11. M. Weinert and A. J. Freeman, J. Magn. Magn. Mater. 38, 23 (1983).
12. E. Tamura and R. Feder, Solid State Commun. 44, 1101 (1982).
13. J. Korecki and U. Gradmann, Phys. Rev. Lett. 55, 2491 (1985).
14. C. L. Fu and A. J. Freeman, to be published.
15. M. Landolt and M. Campagna, Phys. Rev. Lett. 38, 663 (1977).
16. I. P. Moore and J. B. Pendry, J. Phys. C 11, 4615 (1978).
17. C. Rau, J. Magn. Magn. Mater. 30, 141 (1982).
18. R. Feder, J. Phys. C 14, 2049 (1981).
19. C. S. Wang and A. J. Freeman, Phys. Rev. B 21, 4585 (1980).
20. O. Jepsen, J. Madsen, and O. K. Andersen, Phys. Rev. B 26, 2790 (1982).
21. A. J. Freeman, D.-S. Wang, and H. Krakauer, J. Appl. Phys. 53, 1997 (1982).
22. H. Krakauer, A. J. Freeman, and E. Wimmer, Phys. Rev. B 28, 610 (1982).
23. E. Wimmer, A. J. Freeman, and H. Krakauer, Phys. Rev. B 30, 3113 (1984).
24. P. E. Ferguson, J. Appl. Phys. 49, 2203 (1978).
25. G. Allan, Phys. Rev. B 19, 1447 (1979); also see his paper in these proceedings for a historical review and perspective of the work.
26. S. Matsuo and I. Nishida, J. Phys. Soc. Jpn. 49, 1005 (1980).
27. C. Rau and S. Eichner, Phys. Rev. Lett. 47, 939 (1981).
28. F. Meier, D. Pescia, and T. Schriber, Phys. Rev. Lett. 48, 645 (1982).

29. G. Gewinner, J. C. Peruchetti, and A. Jaegle, Phys. Rev. B 27, 3358 (1983).
30. L. E. Klebanoff, S. W. Robey, G. Liu, and D. A. Shirley, Phys. Rev. B 30, 1048 (1984). See also paper by L. E. Klebanoff et al. in these proceedings.
31. D. R. Grempel, Phys. Rev. B 24, 3928 (1981).
32. N. Hirashita, G. Yokoyama, T. Kambara, and K. I. Gondaira, J. Phys. F 11, 2371 (1981).
33. C. L. Fu and A. J. Freeman, Phys. Rev. B 33, 1755 (1986).
34. R. H. Victora and L. M. Falicov, Phys. Rev. B 31, 7335 (1985).
35. H. Hasegawa, ISSP Tech. Rep. A 1585 (1985).
36. C. L. Fu, A. J. Freeman, E. Wimmer, and M. Weinert, Phys. Rev. Lett. 54, 2261 (1985).
37. J. S. Foord, A. P. C. Reed, and R. M. Lambert, Surf. Sci. 129, 79 (1983).
38. G. Zajac, S. D. Bader, and R. J. Friddle, Phys. Rev. B 31, 4947 (1985).
39. R. J. Wilson and A. J. Miller, Surf. Sci. 128, 70 (1980).
40. S. Ohnishi, C. L. Fu, and A. J. Freeman, J. Magn. Magn. Mater. 50, 161 (1985).
41. C. L. Fu, A. J. Freeman, and T. Oguchi, Phys. Rev. Lett. 54, 2700 (1985).
42. C. L. Fu and A. J. Freeman, J. Magn. Magn. Mater. (to be published).
43. T. Oguchi and A. J. Freeman, J. Magn. Magn. Mater. (to be published).

Magnetic and Electronic Properties of Transition-Metal Surfaces and Overlayers

*R.H. Victora**

Department of Physics, University of California, and Materials and Molecular Research Division, Lawrence Berkeley Laboratory, Berkeley, CA 94720, USA

Results of calculations for the magnetic and related electronic properties of transition-metal surfaces and overlayers are presented for a variety of systems. They involve Co, Fe, and Cr in a diversity of forms, including alloys, metastable configurations, and overlayers on nonmagnetic metals. The overall behavior of these systems is interpreted in terms of four qualitative rules which are presented, analyzed, and illustrated.

1. Introduction

There is considerable current interest in the magnetic and related electronic properties of 3d magnetic transition-metal surfaces and overlayers. These metals exhibit itinerant magnetism: Their magnetization derives from the spin polarization of the itinerant d electrons. In moving down the periodic table from Ni, there is a decrease in the number of these d electrons (an increase in the number of d holes), and a consequent increase in the bulk magnetization [1] from 0.61 Bohr magnetons in Ni, to 1.72 in Co, and 2.22 in Fe. Beyond Fe lie the more complicated magnetic structures of Mn and Cr. In particular Cr has an antiferromagnetic ground state [2] in which at the maximum of an incommensurate spin density wave there is a magnetization of 0.59 Bohr magnetons. In all these elements, the itinerant nature of the d electrons makes the magnetic properties a sensitive function of local environment. Consequently the presence of a dissimilar neighbor, as found in an interface, or the absence of some neighbors, as found at a surface, may cause considerable changes in the local magnetic properties.

We have calculated the electronic and magnetic properties for many surface and overlayer systems [3-6]. We used a Slater-Koster parametrized tight-binding scheme in which the one- and two-center integrals are fitted to the bulk band structures of the elements; 4s, 4p and 3d electrons are included. The electron-electron interaction consists of single-site contributions and is sufficiently general to allow for realistic effects such as nonrigid exchange splitting. The interaction is treated self-consistently in the Hartree-Fock approximation. Our scheme (see Table 1) has been tested against experimental data [7,8] and against state-of-the-art first-principles calculations [9,10] on several occasions, and has produced consistently excellent agreement [3-6,11,12].

This contribution provides a summary of our most important predictions for the magnetic and related properties of a variety of surface and overlayer systems. Within each section, background information on the important issues and material properties is presented. Comparisons with experiment and other theories are made, and our most unusual or unexpected results are examined in detail.

Finally, the combination of our results with those of previous workers allows, in the concluding section, the derivation of a set of qualitative rules for the behavior of the magnetization in heterogeneous systems.

* Present address: Kodak Research Laboratories, Rochester, NY 14650.

TABLE 1
Hamiltonian Parameters.

Listed are the 24 Slater-Koster tight-binding matrix elements, the U necessary to produce the correct spin polarization, and the occupancies in the paramagnetic state. A slightly different (wider conduction band) set of bcc Co parameters was used in the Co on Cu calculation. The set listed was used for the FeCo calculations and is believed to be more accurate.

	bcc Cr	bccFe	fcc Fe	bcc Co	fcc Co
$(ss)_0$	.594	.514	.600	.507	.560
$(pp)_0$	1.173	1.118	1.384	1.111	1.344
$(dde_g)_0$	+.005	-.068	-.051	-.076	-.092
$(ddt_{2g})_0$	-.016	-.089	-.063	-.094	-.102
$(ss\sigma)_1$	-.10605	-.10605	-.07452	-.10605	-.07452
$(sp\sigma)_1$	-.19326	-.19326	+.14706	-.19326	+.14706
$(sd\sigma)_1$	-.08740	-.08018	+.08777	-.07537	+.08209
$(pp\sigma)_1$	+.26932	+.26932	+.30843	+.26932	+.30843
$(pp\pi)_1$	-.00672	-.00672	-.01129	-.00672	-.01129
$(pd\sigma)_1$	.10807	+.09915	+.11339	+.09320	+.10605
$(pd\pi)_1$	-.02689	-.02467	-.02049	-.02319	-.01917
$(dd\sigma)_1$	-.06735	-.05669	-.05217	-.05009	-.04563
$(dd\pi)_1$	+.04441	+.03738	+.02647	+.03303	+.02315
$(dd\delta)_1$	-.00795	-.00669	-.00290	-.00591	-.00254
$(ss\sigma)_2$	-.05276	-.05276	-.05495	-.05276	-.05495
$(sp\sigma)_2$	-.06725	-.06725	+.08064	-.06725	+.08064
$(sd\sigma)_2$	-.03308	-.03035	.00000	-.02853	.00000
$(pp\sigma)_2$	.17199	+.17199	+.15017	+.17199	+.15017
$(pp\pi)_2$	.03242	+.03242	-.05749	+.03242	-.05749
$(pd\sigma)_2$	.05670	+.05202	.00000	+.04890	.00000
$(pd\pi)_2$	-.00100	-.00092	.00000	-.00086	.00000
$(dd\sigma)_2$	-.04000	-.03367	.00000	-.02975	.00000
$(dd\pi)_2$	.01188	+.01000	.00000	+.00884	.00000
$(dd\delta)_2$	-.00089	-.00075	.00000	-.00066	.00000
U	.167	.220	.220	.296	.296
n^0_{sp}	1.175	1.185	1.154	1.174	1.150
n^0_{t2g}	3.258	4.223	3.954	4.478	4.448
n^0_{eg}	1.567	2.592	2.892	3.348	3.404
n^0_{total}	6.000	8.000	8.000	9.000	9.000

2. Surfaces of Ferromagnetic Iron and the Ferromagnetic Ordered Iron-Cobalt Alloy

The effort to understand surface behavior must start with an understanding of the bulk material. Iron is an element of great technological importance and its properties have been extensively examined. For our purposes, it is sufficient to note that Fe possesses a *g* factor [1] of approximately 2.10 and thus a spin polarization of 2.12. This means that Fe is ferromagnetically weak, i.e., not all available d holes contribute to the magnetization. This has been shown [11] to be a direct consequence of a relatively weak electron-electron interaction.

The Fe-Co alloy is also technologically important [13]. This alloy possesses several useful characteristics such as the largest known magnetization per atom and an extremely high Curie temperature. In particular, the ordered FeCo alloy, which consists of Fe and Co atoms arranged at the corner and body-centered positions of a simple-cubic lattice, has a magnetization of 2.42 μ_B per atom [14]. This is 0.45 μ_B higher than the average of its constituent elements. Neutron-diffraction studies [15] indicate that the vast majority of the anomalous increase in the magnetization is caused by an increase in the Fe magnetic moment from 2.2 μ_B to approximately 3.0 μ_B, while the Co magnetic moment increases only slightly. MEYER and ASCH [16] found the g factor of the equiatomic alloy to be approximately 2.15, thus suggesting that the spin polarization is 2.25 ± 0.01. Victora and Falicov demonstrated [11] that, at least for the disordered equiatomic alloy, the increased magnetization is caused by magnetic saturation made possible by the presence of the strong Co electron-electron interaction.

A wide variety of experimental techniques [7-8,17-22] show that the surface layer of a magnetic 3d transition metal is magnetic, while recent theoretical studies demonstrate that the surface magnetization is, in fact, enhanced relative to the bulk value [10,12,23]. In particular, theory predicts a spin polarization of 2.98 electrons for the Fe(100) surface [10] and approximately 0.70 electrons for the Ni surfaces [12,23]. The experimental evidence of GRADMANN et al. [7] suggests that the magnetization of the Fe(110) surface is enhanced by approximately 30%. It is found [24] that Fe atoms segregate to the surface of the Fe-Co alloy.

Our calculation gives a surface spin polarization for the Fe(100) surface of 2.90 electrons. The enhanced magnetization penetrates some distance into the bulk: The second, third, and center layers have spin polarizations of 2.30, 2.24, and 2.18 electrons, respectively. These results are very similar to those of OHNISHI et al. [10]. In particular, their surface spin polarization is 2.98 electrons and their calculated DOSs also resemble ours, although 5-10% narrower. This excellent agreement between two different calculational techniques confirms the accuracy of our methods. We find that the spin polarization also increases at the Fe(110) surface, where it reaches a value of 2.55. This is in agreement with the experimental results of GRADMANN et al. [7]. The relative magnetizations of bulk Fe and the two surfaces are easily explained by noting that nearest-neighbor interactions dominate the d-band width (matrix elements are proportional [25] to $(\text{distance})^{-5}$ and the first-nearest-neighbor distance is $(1/2)\sqrt{3}$ smaller than the second-nearest-neighbor (distance)). The bulk atom in the bcc structure has eight nearest neighbors, an atom on the (110) surface has six, and an atom on the (100) surface has four. Thus the (100) surface has the fewest neighbors, smallest bandwidth, the same electron-electron interaction as the other geometries, and consequently the highest magnetization. The atom on the (110) surface has an intermediate number of neighbors and consequently will have an intermediate magnetization.

A necessary prerequisite to the study of the FeCo surface properties is a proper understanding of the bulk material itself. We calculate the spin polarization of bulk FeCo to be 4.44 electrons per unit cell, thus capturing more than 90% of the experimental enhancement observed in the alloy relative to the average of the pure elemental magnetization. Most of the increase is found to occur on the Fe atom, where the presence of eight neighboring Co atoms helps increase the effective electron-electron interaction and saturates the moment. The much smaller magnetic enhancement found on the Co atom is caused by the shift from fcc to bcc structures and possibly a slight swelling in the lattice constant as iron is added.

The simple-cubic FeCo structure may be cut by the (100) plane, leaving either all Fe atoms at the surface or all Co atoms. The DOS of the Fe surface bears a strong resemblance to the DOS at the elemental Fe(100) surface. The spin polarization is 2.95 electrons, which is also very similar to the Fe(100) result of 2.90 electrons. These results suggest that the presence of the (100) surface is the dominant effect and that the precise nature of the second layer is not too important for the surface Fe layer. This is reasonable since, as previously observed, the effect of Co is to increase the effective electron-electron-interaction to bandwidth ratio at the Fe atom and it is clear from Fe(100) results that a (100) surface can do this much more effectively. Consequently, the presence of Co instead of Fe is somewhat unimportant. The one major difference is that the surface effects are not able to penetrate even one layer deep into the bulk. This is presumably because Co, unlike Fe, has a saturated magnetization. Consequently, spin polarization and related properties are less susceptible to perturbation.

The all-Co surface has slightly larger exchange splitting and spin polarization than the bulk (2.03 vs. 1.78 electrons). Interestingly, there is a very slight (0.1 electron) decrease in the spin polarization of the underlying Fe layer. However, even this effect vanishes at the next Co layer and deeper. Overall, the surface behaves like Ni surfaces [12,23] in that the surface DOS is narrowed and new features appear, but the magnetic saturation allows only small increases in the magnetization.

The (110) surface of FeCo contains both Fe and Co atoms arranged in a two-atom unit surface cell. As expected from the nearest-neighbor arguments given previously, the (110) surface displays behavior intermediate between bulk FeCo and the (100) FeCo surfaces. The calculated spin polarizations at the surface are 2.75 electrons for Fe and 1.86 electrons for Co. There is very little penetration of surface effects into the bulk.

Experimental and theoretical studies indicate that Fe tends to segregate to the surface of the FeCo alloy. To represent this, we have calculated the electronic properties of a Fe(110) overlayer, one atom thick, atop the FeCo alloy. The spin polarizations are 2.63 electrons for the Fe atom occupying the Co position and 2.67 electrons for the Fe atom occupying the normal position. Clearly, the two Fe atoms have very similar electronic properties. It is to be noted that their properties are intermediate between the Fe(110) surface and the FeCo(110) surface.

This section has examined two Fe surfaces, bulk FeCo, three FeCo surfaces, and the segregated surface. These systems have served to exemplify two very important ideas: (1) Surfaces tend to increase magnetization; and (2) Co is magnetically saturated; Fe is magnetically weak. These two ideas will reappear in later sections and, finally, be made more specific in the conclusion.

3. Cobalt Overlayers on Copper (111)

Recently, there has been renewed interest in the magnetism and related properties of thin magnetic transition-metal layers deposited on non-magnetic substrates. Experimental evidence shows that two layers of Ni on Cu form a magnetic system [22,26], but Ni on a Pb-Bi alloy substrate or on an Al substrate is paramagnetic below 2.5-3 atomic layers [27,28]. However, Co and Fe retained their magnetic moment on these same substrates, even when deposited in subatomic layers [27,28]. Theoretical investigation shows that one layer of Ni on Cu(100) is substantially magnetic [9,12] while the same system on the (111) surface is not [12]. As discussed in the previous section, the effect of a surface is to increase the magnetization. This suggests that any possible reduction in magnetization is caused by the substrate. In fact, it is believed [12] that the crucial mechanism acting to suppress Ni magnetization at the Ni-Cu interface is hybridization of the Ni d band with the Cu sp band, which changes the shape of the band edge and reduces the "effective" number of d holes. Both Fe and Co, which have many more holes, should be relatively immune to this effect.

The particular system addressed in this section [Co on Cu(111)] has also been studied previously. GONZALEZ et al. [29] and MIRANDA et al. [30] used angle-resolved photoemission spectroscopy to determine a surface density of states for one and two layers of Co deposited on a clean, well-ordered Cu(111) surface. Observations at the surface Brillouin-zone center produced several large peaks, one of which is near the Fermi energy and thus not normally associated with the Cu density of states (DOS). The similarity between this peak and a bulk Co DOS, interpreted as ferromagnetic by HIMPSEL and EASTMAN [31], suggested to them that Co is magnetic with an exchange splitting of 0.7 eV at the $\bar{\Gamma}$ point of the surface Brillouin zone. Further support for the ferromagnetism of this system arises from Mossbauer spectroscopy which shows that two and four layers of Co are magnetic when epitaxially grown on Cu(111) surfaces [32].

In this section I present results of calculations for the magnetic properties of thin (one- and two-layer) Co films deposited on the Cu(111) surface. Calculations are performed using both a one-atom hexagonal unit surface cell (symmetry C_{3v}) and a two-atom rectangular unit cell. This latter cell permits possible breaking of spatial symmetry such as antiferromagnetism, ferrimagnetism, etc.

Our calculations show that for one layer Co on the Cu(111) surface, the spin polarization of the Co atom equals 1.63. However, approximations within the calculation may slightly exaggerate this polarization which suggests that the true value is essentially unchanged from the bulk value: 1.56. This is in interesting contrast to the case of Ni on the Cu(111) surface in which the Ni magnetization is greatly suppressed. Of course, there is no contradiction present since previous calculations [12] have shown that small changes in substrate orientation or material composition can produce large effects. This is attributable to the delicate balance between two competing effects: enhancement by the surface and suppression by the substrate hybridization.

The d orbitals of the Co layer can be classified uniquely by specifying their bulk and surface symmetries. This facilitates detailed examination of the distribution of magnetization among orbitals. One finds that the orbital of symmetry Λ_1 has significantly less spin polarization and exchange splitting than the orbitals of symmetry Λ_3. This is presumably due to its geometric orientation perpendicular to the (111) plane, which means that it points directly into the Cu substrate. In addition, its symmetry is the same as most of the Cu sp electrons, which facilitates hybridization. One interesting conclusion to be drawn is that rigid-band exchange splitting is only approximately obeyed in this Co-on-Cu(111) system. This contrasts to the bulk system, where our calculations found only negligible (less than 1%) exchange-splitting difference between orbitals.

We also performed our calculations for two layers of Co-on-Cu(111). The top and second layers have spin polarizations of 1.65 and 1.58, respectively. The spin polarization is distributed much more evenly between orbitals although, relative to the Λ_3 states, the Λ_1 orbital has a slightly enhanced (~0.06) value in the surface layer and a slightly suppressed (~0.05) value in the second layer. The Cu atoms have negligible spin polarization (~0.01) independent of the number of Co overlayers.

Agreement between our theory and the experiment of Miranda and co-workers is considerable, but discrepancies do exist. The quantitative correspondence between theoretical and experimental peak locations for one layer Co-on-Cu is excellent. The photoemission spectra at the $\bar{K}$ point for both this system and for the clean Cu(111) surface can be consistently explained by theoretical states which transform according to Λ_1, the identity representation. More disappointing is the experimentally observed absence of the theoretically predicted shift in peak locations as an additional layer of Co is added. The experimentally observed displacement in Co peak location with photon energy is also puzzling, since monolayer states can have no dispersion.

Through the employment of the two-atom rectangular unit surface cell discussed previously, we have found three locally stable configurations for the system consisting of one layer of Co-on-Cu(111). The simplest of these is the ferromagnetic arrangement discussed earlier. In this state both of the atoms in the unit cell possess equal polarizations and are identical in all respects.

We also examine an antiferromagnetic state in which the spin polarizations of the two atoms in the unit cell have equal magnitude (1.51 per atom) but opposite orientation. Those states which possess a positive reflection symmetry are found to have the lowest spin polarization, presumably due to their hybridization with the Cu s states. The total energy of this configuration is found to be considerably higher (0.15 Ry per surface atom) than that of the ferromagnetic system, which implies that it is not the ground state.

Our third configuration is a new kind of state which we call "spatially modulated." This state possesses an approximately equal polarization of 1.60 on

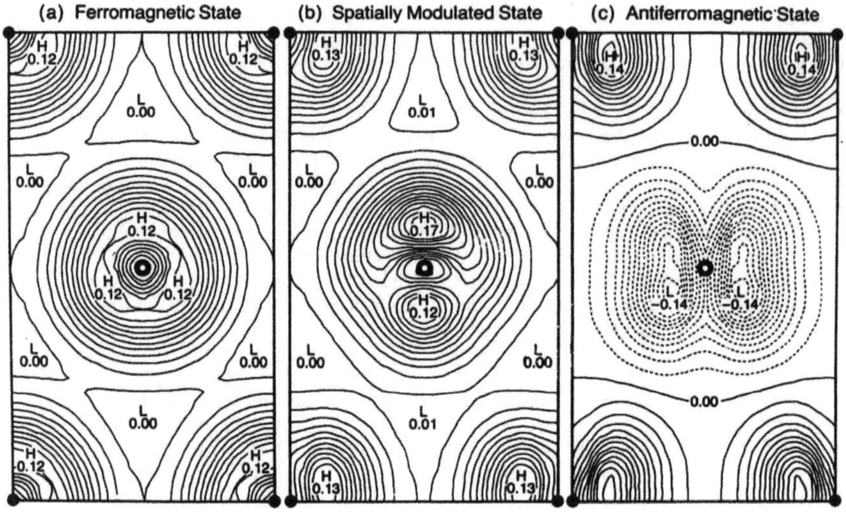

Fig. 1. Real-space magnetization density of a Co monolayer on Cu(111): the Co surface layer for (a) ferromagnetic state; (b) spatially modulated state; (c) antiferromagnetic state. Spin-up density contours are plotted in full lines; spin-down density contours in dashed lines. All graphs are projections on the Co(111) plane of the magnetization of the "vacuum" half-space.

each atom of the unit cell, but it is distributed differently between the orbitals. The first three orbitals, which form the t_{2g} representation in the bulk, have equal polarization on both atoms; but the last two orbitals, the "e_g orbitals," switch the polarization between them as one moves from one atom to the other. This redistribution in space leads to the expression "spatially modulated." It is not to be confused with ferrimagnetic states which possess opposite and unequal spins on the two atoms of the unit cell. The total energy of the spatially modulated state is calculated to be 0.03 Ry per surface atom above the ferromagnetic state. Distinctions of this small magnitude are probably not within the accuracy of our approximations. Consequently, we conclude that either the ferromagnetic or spatially modulated state could be the ground state.

Figure 1 illustrates the spatial distribution of magnetization. The plots were calculated by multiplying the spin polarization by the correct spherical harmonic and an atomic radial function. Most trivially, the figures exhibit the C_{3v} symmetry of the ferromagnetic configuration (one reflection line is vertical, the other two at 120° to it) and the single vertical reflection line of the other two configurations. The wavelike modulation of the spin polarization in the antiferromagnetic case is also clearly visible. More interesting is the asymmetric accumulation of magnetization density above the middle Co atom in the spatially modulated state, leading to the concept of a small wave of periodicity equal to two atoms, superposed on a more constant background.

In concluding this section, attention is drawn to three distinct results: (1) Both one and two layers of Co on a Cu(111) surface are magnetized with a spin polarization close to the bulk value. (2) There exists a new kind of broken symmetry state which we have called spatially modulated. (3) Interface magnetization is reduced by the paramagnetic substrate. This latter result will reappear in the concluding section.

4. Chromium Surfaces and Chromium Monolayers on Iron

Bulk chromium occurs in the body-centered cubic structure with an antiferromagnetic (AF) ground state [2] modulated by an incommensurate spin density wave

(SDW). The SDW is in one of the (100) directions with a wavelength of approximately 21 lattice spacings. The magnetization at the maximum is 0.59 Bohr magnetons. Experimentally it is found that the addition of small amounts (approximately 1%) of group VII impurities such as Mn produces a simple AF structure with a magnetic moment of approximately 0.67 Bohr magnetons. In this structure atoms in the body-centered positions of the bcc lattice have spins pointing only in on direction; atoms in the corner positions have spins in the opposite direction.

This simple AF structure demands that (100) planes contain atoms of only one spin direction. Consequently, the (100) surface is expected to possess ferromagnetic order. Evidence of this planar ferromagnetism is found in the electron-capture spectroscopy results of RAU and EICHNER [8]. Their sample, however, displayed a c(2x2) structure indicative of impurities, which may have affected their results. The (100) surface has been examined also by two angle-resolved photoemission experiments [34,35]; both find a surface state or resonance at an energy approximately 0.70 eV below the Fermi level. The measured temperature dependence of this resonance in the latter experiment is taken as evidence for a ferromagnetic surface [35].

The (100) surface of chromium has also been examined theoretically [36,37] by means of a simple exchange interaction and a tight-binding approach which neglected the sp conduction electrons. ALLAN [36] finds that the surface magnetization is very large (2.8 Bohr magnetons) relative to the bulk and argues that this enhancement should penetrate into the bulk, decreasing by a factor (-0.5) per layer. (The negative sign refers to the AF.) GREMPEL [37] finds a surface spin polarization of 2.6 Bohr magnetons and a very high surface Curie temperature. FU, FREEMAN and OGUCHI, in a very recent paper [38], cite a value of 2.49 μ_B for the surface magnetization.

The (110) planes cut the simple AF bcc lattice so that an equal number of up and down spins are encountered. The (110) surface should therefore be an AF one. There have been two photoemission experiments on the Cr(110) surface. The experiment of JOHANSSON et al. [39] found no evidence of surface states; the later experiment of WINCOTT et al. [40] measured the dispersion of a surface state (binding energy of approximately 0.2 eV) along $\bar{\Sigma}$. The periodicity of their spectra suggests an AF surface.

Stainless steel is the name given to a family of iron based alloys which contain at least 12% chromium. Auger electron spectroscopy [41] shows that when a 304 stainless steel sample (a common type) is heated, there is a strong enhancement of the Cr concentration at the surface. This is in agreement with chromium's lower heat of crystal formation [42]. It is this enhanced concentration of Cr at the surface which, when oxidized, is influential in preventing the rusting of the steel.

We find the surface spin polarization at the Cr(100) surface to be 3.00 electrons. This enhancement relative to the bulk, a factor of 5.1 from the SDW maximum, is much larger than that [4,12] found in Fe and Ni. The magnitude of the enhancement can be attributed to the large number of unpolarized d-holes present in the bulk. Consequently, the decreased bandwidth found at the Cr surface, which leads to a stronger effective magnetic interaction, can greatly increase the surface spin polarization. Elements like Fe or Ni, with fewer available unpolarized holes, experience smaller changes in the same local environment.

The narrowing of the Cr DOS at the (100) surface is shown in Figure 2. It is clear that most d-holes occur in the minority subband. This subband is essentially concentrated in a single peak structure entirely above the Fermi level. The surface DOS can be compared with the bulk DOS given in [5]. There the d-holes are almost evenly distributed between the two subbands. Furthermore, it is clear that both subbands have sizable strength on either side of the Fermi energy, which falls in a valley of the bcc DOS.

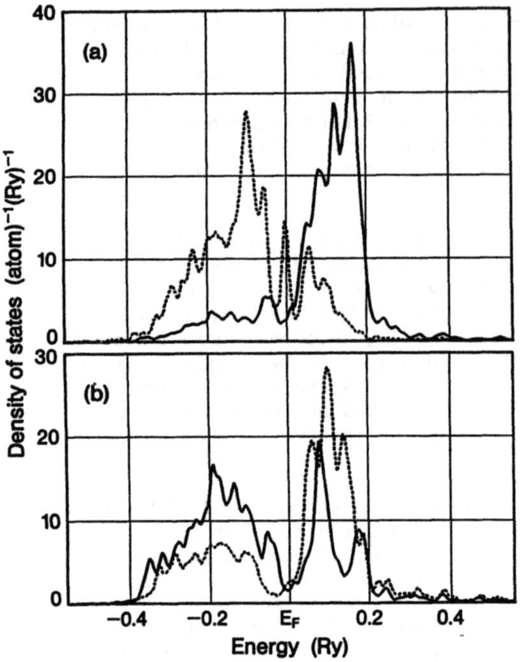

Fig. 2. The d-orbital component of the projected density of states. (a) The chromium (100) surface layer. (b) The chromium (100) second layer. Solid lines are states with the spin orientation of the surface minority states; dashed lines correspond to the majority states.

Figure 2(b) shows the DOS projected at the layer immediately below the surface. The width of each subband is approximately equal to the bulk value, which is consistent with the presence of all nearest, and most second-nearest, neighbors. The spin polarization is opposite to that of the surface layer, which is consistent with the AF of chromium. However, it is clear that there is a substantial difference between the minority and the majority DOS, suggesting that the spin polarization is not bulklike.

The spin polarization of the second layer is (-1.56), substantially different from the bulk. The spin polarizations of the third through sixth layers are 1.00, (-0.93), 0.86, and (-0.85), respectively. A similar penetration of the enhanced surface magnetization was predicted in Section 2 for the Fe(100) surface, although there the effect is much smaller. This penetration appears to be a direct consequence of the easy availability of unpolarized d-holes in bulk Cr. Each atoms feels the larger exchange splitting of its neighbor towards the surface and responds by increasing its own; this is, in the case of Cr, an energetically very inexpensive process.

We find a very large pair of surface states at the Γ point. Both their symmetry and dispersion characteristics are in agreement with that feature found by KLEBANOFF et al. [35]. Furthermore, we find the states to be spin-polarized. This then supports the experimentally proposed arguments [35] for the Cr(100) surface ferromagnetism.

Our (110) surface calculation gives a two-atom unit surface cell with AF ordering as the ground-state configuration. We also find a surface feature with AF periodicity which disperses similarly to that found by WINCOTT et al. [40], thus reinforcing their experimental deduction of AF surface ordering. This ordering is in agreement with a simple saw cut of the bulk along the (110) plane. We were unable to find a ferromagnetic locally stable minimum in the ground-state energy.

The surface spin polarization in the AF configuration is 2.31 electrons. This value is smaller than that for the (100) surface as one would expect, since the

(110) surface atom has six nearest neighbors, as opposed to four in the (100) surface. The spin polarization in the second layer, 1.00 electrons, is considerably smaller than that of a second-layer (100) atom, 1.56. This is a consequence not only of the smaller (110) surface perturbation, but also of the larger interlayer distance in the (110) direction. In particular, (100) layers are separated by 0.5 a (where a is the lattice constant), whereas (110) layers are separated by 0.707 a, with a consequent smaller coupling between layers.

We have calculated the spin polarization of a Cr monolayer atop the Fe(100) surface to be 3.63 electrons, with a ferromagnetic arrangement pointing oppositely to the underlying Fe substrate. This surprising prediction has been supported by the recent results of FU et al. [38] who also predict a large monolayer polarization (3.1 electrons). It is interesting that this large magnetization occurs for a system as important as the pre-oxidized stainless-steel surface. We were unable to obtain a local stable minimum for a Cr ferromagnetic layer, ferromagnetically ordered with respect to the iron.

Insight into this result may be gained by comparing it with the dilute FeCr alloy. Neutron scattering [43] results show that the isolated Cr atoms point oppositely to the surrounding Fe bulk and have a spin polarization of 1.2 electrons, twice the maximum bulk value of the SDW. This latter result presumably stems from the stronger electron-electron interaction in Fe, and a consequent stronger exchange splitting which helps Cr increase its own splitting and magnetization. This is the same argument explored in great detail for the iron-cobalt alloy [4], where Co has the stronger electron-electron interaction. The combined effect of diminished number of neighbors and stronger Fe exchange results in the calculated large Cr moment. We may understand the AF coupling by noting that Mn is the element intermediate between Fe and Cr, which suggests that the Fe-Cr interaction may be similar to the Mn-Mn interaction. Manganese exhibits localized moments but no ferromagnetism, implying that Fe and Cr would not couple ferromagnetically either.

Consideration of a Cr monolayer atop the Fe(110) surface suggests several alternative configurations. One might expect the monolayer to be AF because of the AF interaction of each Cr atom with its four Cr neighbors. This arrangement forces half the Cr atoms to have a ferromagnetic interaction with the four underlying nearest Fe atoms. As already discussed, Fe has a very strong effect on the Cr spin polarization (it doubles the value of the Cr polarization in the dilute alloy). The Fe substrate therefore would favor a ferromagnetic Cr monolayer, with AF order with respect to the bulk. Clearly there are two competing effects, and the nature of the ground state can be determined only after a detailed calculation.

We find that the ground state consists of a ferromagnetic Cr monolayer with its spins oriented in the direction opposite to the Fe substrate, similar to the Cr on Fe(100) arrangement. The spin polarization of the Cr is 2.25 electrons, smaller than the pure Cr surface. There is, as in the (100) monolayer, a small decrease in the spin polarization of the underlying Fe layer to a value of 2.03 electrons. Changes in the Cr monolayer DOS relative to the bulk DOS are not spectacular and mostly they reflect just the increased spin polarization.

We find that a structure consisting of an AF chromium layer is metastable: It produces a local minimum in the total energy curve, 0.05 Ry/surface atom above the ground state. The spin polarization of the two different Cr atoms are 3.03 and (-3.31) electrons, with the larger magnitude corresponding to the atom with AF arrangement to both its Cr and Fe nearest neighbors. It is clear that because of the different magnitudes of the spin polarizations the Cr monolayer is not truly AF, but rather ferrimagnetic.

5. Conclusion

This paper has described calculations for the magnetic and related electronic properties of a variety of transition metal systems (see Table 2). Some of the

Table 2

Spin-Polarization Results for All Ordered Systems

System	Spin Polarization
fcc Co	1.56
bcc Co	1.80
bcc Fe	2.12
bcc Cr	0.67
sc FeCo - Fe atom	2.66
sc FeCo - Co atom	1.78
(100) Fe surface	2.90
(110) Fe surface	2.55
(100) Cr surface - surface layer	3.00
(100) Cr surface - second layer	-1.56
(110) Cr surface	2.31
(100) FeCo surfaces - Fe surface layer	2.95
(100) FeCo surfaces - Co surface layer	2.03
(110) FeCo surface - Fe atom	2.75
(110) FeCo surface - Co atom	1.86
(110) Fe monolayer on FeCo - Co position	2.63
(110) Fe monolayer on FeCo - Fe position	2.67
(100) Cr monolayer on Fe	3.63
(110) Cr monolayer on Fe	2.25
(111) Co monolayer on Cu	1.63
(111) Co dilayer on Cu - surface layer	1.65
(111) Co dilayer on Cu - interface layer	1.58

more unusual results were the huge magnetizations found in the Cr(100) systems and the possibility of a new kind of magnetic order in the Co monolayer on Cu. The large variety of our results also suggests the possibility of a set of summarizing rules.

Four of the calculations have been made for pure transition metal surfaces. These calculations [Cr(100), Cr(110), Fe(100), and Fe(110)] all demonstrate that the removal of neighboring atoms decreases the projected bandwidth on the surface. Consequently, the electron-electron interaction to bandwidth ratio increases and so does the magnetism. In short, removal of the nearest neighbor atoms of its own kind tends to enhance magnetism.

These four surface calculations also help justify the next rule. It is recalled that the Cr surface enhancement of the magnetization is enormous; four to five times over bulk values. The enhancement at the Fe surface is smaller (20% to 40%) but still significant. The enhancement at the FeCo surfaces is quite small: less than 15%. The relevant difference between these systems is the number of available unpolarized d holes. FeCo has very few; Fe has approximately 0.7 per atom, and Cr has more than 4 per atom. (The concept of d-holes is not entirely concrete due to the presence of sp-d hybridization; it is the qualitative idea that is important.) Consequently, magnetization enhancement is most pronounced in those elements where the bulk magnetization is not close to saturation.

Bulk FeCo displays an interesting principle in itself. Section 2 argued that the strongly exchanged split Co can help increase the Fe magnetization. This same effect occurs in the Cr near surface region where the strongly exchange split surface Cr atoms help increase the spin polarization of the lower Cr layers. In general, we conclude that the presence of a strongly magnetized atom with a large exchange splitting near a weakly magnetic but polarizable atom with a smaller splitting considerably enhances the magnetization of the latter.

The final rule is exemplified by the calculations for the Co overlayers on Cu, in which it is found that a Co surface layer in contact with Cu has a lower spin polarization than a surface layer only in contact with Co. This suppression of magnetization by a nonmagnetic substrate has been explored in a number of experimental and theoretical studies as described in the introduction to Section 3. Examination of all these results suggests that the presence of a nonmagnetic unpolarizable atom next to - and coupled to - a magnetic transition metal element tends to decrease the magnetization of the latter.

It is hoped that these four qualitative rules bring some additional order to the systematics of heterogeneous systems displaying itinerant magnetism.

Acknowledgments

All calculational results reviewed in this paper were originally derived while the author was at U.C. Berkeley, collaborating with L. M. Falicov. The work was supported at the Lawrence Berkeley Laboratory by the Director, Office of Energy Research, Office of Basic Energy Sciences, Materials Sciences Division of the Department of Energy under Contract No. DE-AC03-76SF00098. I would also like to acknowledge an A.T.&T. Bell Laboratories predoctoral fellowship, under whose sponsorship much of this work was done.

References

1. C. Kittel, Introduction to Solid State Physics, 5th ed. (Wiley, New York, 1976).
2. S. A. Werner, A. S. Arrott, H. Kendrick, Phys. Rev. 155, 528 (1967); G. Shirane, W. J. Takei, J. Phys. Soc. Jpn. Supp. 17, B III, 35 (1962); W. C. Koehler, R. M. Moon, A. L. Trego, and A. R. MacKintosh, Phys. Rev. 151, 405 (1966).
3. R. H. Victora and L. M. Falicov, Phys. Rev. B 28, 5232 (1983).
4. R. H. Victora, L. M. Falicov, and S. Ishida, Phys. Rev. B 30, 3896 (1984).
5. R. H. Victora and L. M. Falicov, Phys. Rev. B 31, 7335 (1985).
6. L. E. Klebanoff, R. H. Victora, L. M. Falicov, and D. A. Shirley, Phys. Rev. B 32, 1997 (1985).
7. U. Gradmann, G. Waller, R. Feder, and E. Tamura, J. Magn. Magn. Mat. 31-34, 883 (1983).
8. C. Rau and S. Eichner, Phys. Rev. Lett. 47, 939 (1981).
9. D. S. Wang, A. J. Freeman, and H. Krakauer, Phys. Rev. B 26, 1340 (1982).
10. S. Ohnishi, A. J. Freeman, and M. Weinert, Phys. Rev. B 28, 6741 (1983).
11. R. H. Victora and L. M. Falicov, Phys. Rev. B 30, 259 (1984).
12. J. Tersoff and L. M. Falicov, Phys. Rev. B 26, 6186 (1982).
13. G. Couderchon and J. F. Tiers, J. Magn. Magn. Mat. 26, 196 (1982).
14. D. I. Bardos, J. Appl. Phys. 40, 1371 (1969).
15. M. F. Collins and J. B. Forsyth, Philos. Mag. 8, 401 (1963).
16. A. J. P. Meyer and G. Asch, J. Appl. Phys. 32, 330S (1961).
17. E. W. Plummer and W. Eberhardt, Phys. Rev. B 20, 1444 (1979).
18. G. P. Felcher, S. D. Bader, R. J. Celotta, D. T. Pierce, and G. C. Wang, in Ordering in Two Dimensions, ed. by S. K. Sinha (Elsevier, New York, 1980), p. 107.
19. W. Eib and S. F. Alvarado, Phys. Rev. Lett. 37, 444 (1976).
20. W. Göpel, Surf. Sci. 85, 400 (1979).
21. A. M. Turner, Y. J. Chang, and J. L. Erskine, Phys. Rev. Lett. 48, 348 (1982).
22. C. Rau, Comments Solid State Phys. 9, 177 (1980).
23. H. Krakauer, A. J. Freeman, and E. Wimmer, Phys. Rev. B 28, 610 (1983).
24. J. L. Morán-López and H. Wise, Appl. Surf. Sci. 4, 93 (1980).
25. W. A. Harrison and S. Froyen, Phys. Rev. B 21, 3214 (1980).
26. D. T. Pierce and H. C. Siegmann, Phys. Rev. B 9, 4035 (1974).
27. G. Bergmann, Phys. Rev. Lett. 41, 264 (1978); Phys. Today 32 (8), 25 (1979).
28. R. Meservey, P. M. Tedrow, and V. R. Kalvey, Solid State Comm. 36, 969 (1980).
29. L. González, R. Miranda, M. Salmerón, J. A. Vergés, and F. Yndurâin, Phys. Rev. B 24, 3245 (1981).

30. R. Miranda, F. Ynduráin, D. Chandesris, J. Lecante, and Y. Petroff, Phys. Rev. B 25, 527 (1982); Surf. Sci. 117, 319 (1982).
31. F. J. Himpsel and D. E. Eastman, Phys. Rev. B 21, 3207 (1980).
32. U. Gradmann, K. Ullrich, J. Pebler, and K. Schmidt, J. Magn. Magn. Mat. 5, 339 (1977).
33. F. Batallan, I. Rosenman, and C. B. Summers, Phys. Rev. B 11, 545 (1975).
34. G. Gewinner, J. C. Peruchetti, A. Jaéglé, and R. Pinchaux, Phys. Rev. B 27, 3358 (1983).
35. L. E. Klebanoff, S. W. Robey, G. Liu, and D. A. Shirley, Phys. Rev. B 30, 1048 (1984).
36. G. Allan, Surf. Sci. 74, 79 (1978); Phys. Rev. B 19, 4774 (1979).
37. D. R. Grempel, Phys. Rev. B 24, 3928 (1981).
38. C. L. Fu, A. J. Freeman, and T. Oguchi, Phys. Rev. Lett. 54, 2700 (1985).
39. L. I. Johansson, L. G. Petersson, K. F. Berggren, and J. W. Allen, Phys. Rev. B 22, 3294 (1980).
40. P. L. Wincott, N. B. Brookes, D. S. Law, G. Thornton, and H. A. Padmore, Vacuum 33, 815 (1983).
41. G. Betz, G. K. Wehner, L. Toth, and A. Joshi, J. Appl. Phys. 45, 5312 (1974).
42. CRC Handbook of Chemistry and Physics, 53rd ed. (CRC Press, Cleveland, 1972), p. F-188.
43. A. T. Alfred, B. D. Rainford, J. S. Kouvel, and T. J. Hicks, Phys. Rev. B 14, 228 (1976).

Magnetic and Superconducting Properties of Surfaces and Thin Films of Ferromagnetic Superconductors

A. Kotani and K. Okada

Department of Physics, Faculty of Science, Osaka University, Toyonaka 560, Japan

1. Introduction.

Since the discovery of magnetic superconductors, much interest has developed in the interplay between magnetism and superconductivity [1]. Especially for bulk systems of ferromagnetic superconductors, $ErRh_4B_4$ and $HoMo_6S_8$, extensive studies have been carried out both theoretically and experimentally [1]. However, for surfaces and films, only a few investigations have been made [2]. The purpose of the present paper is to predict theoretically unusual properties of surfaces and thin films of ferromagnetic superconductors.

In magnetic superconductors, the magnetism is due to the magnetic moments (denoted hereafter by spins) of the 4f electrons of a rare-earth element, whereas the persistent current is carried mainly by the 4d electrons of a transition element. It has been well established that in the ferromagnetic superconductors the electromagnetic interaction between the spin and the persistent current plays an essentially important role in the interplay between magnetism and superconductivity [3,4]. In this paper we disregard the effect of the exchange interaction between the rare-earth spins and the superconducting electrons, which is less important than the electromagnetic interaction.

2. Bulk System.

Before discussing the effect of surfaces, we present a brief survey of the bulk properties of the ferromagnetic superconductors. In $ErRh_4B_4$, for instance, the bulk crystal becomes superconducting below an upper critical temperature T_{c1} which is about 8.7 K, as shown in Figure 1. With decreasing temperature, the rare-earth spins order sinusoidally with a wave number of about 100 $Å^{-1}$ in a narrow temperature range around 1 K [1]. This phase is denoted by the spin-sinusoidal phase, and it coexists with superconductivity. Below a lower critical temperature T_{c2}, the ferromagnetic normal state occurs. The origin of the spin-sinusoidal order is the electromagnetic interaction, by which the persistent current $\vec{J}(\vec{r})$ is induced around each spin and screens the interaction between spins [3,4]. The current $\vec{J}(\vec{r})$ is expressed as

$$\vec{J}(\vec{r}) = -\frac{c}{4\pi\lambda_L^2}\int d\vec{r}\,'\ K(\vec{r}-\vec{r}\,')\ \vec{A}(\vec{r}\,') \ , \qquad (1)$$

where $\vec{A}(\vec{r})$ is the vector potential, λ_L is the London penetration depth, and $K(\vec{r})$ represents the nonlocal effect of the current [$K(\vec{r}) \sim \exp(-r^2/2\xi^2)$ with the coherence length ξ]. The magnetic field $\vec{A}(\vec{r})$ which screens the interaction between spins is associated with $\vec{J}(\vec{r})$ through Maxwell's equation

$$\vec{\nabla} \times \vec{A}(\vec{r}) = \frac{4\pi}{c}\,\vec{J}(\vec{r}) \ . \qquad (2)$$

By means of the relation

$$\vec{\nabla} \times \vec{A}(\vec{r}) = \vec{H}(\vec{r}) + 4\pi\,\vec{M}(\vec{r}) \ , \qquad (3)$$

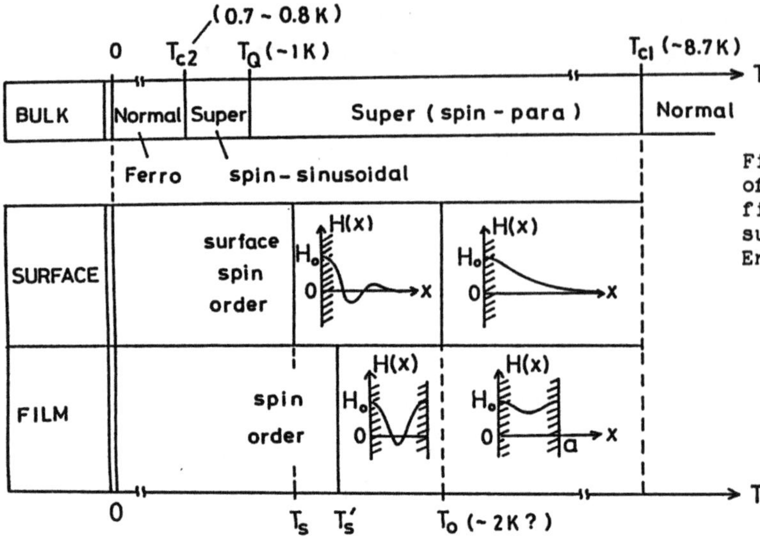

Figure 1. Properties of bulk, surface, and film ferromagnetic superconductors (with $ErRh_4B_4$ as an example).

where $\vec{M}(\vec{r})$ is the magnetization of the rare-earth spins, we eliminate $\vec{J}$ and $\vec{A}$ from (1) - (3), and obtain the relation between $\vec{H}$ and $\vec{M}$ in the form

$$\vec{H}(\vec{q}) = - \frac{4\pi\, K(\vec{q})\, \vec{M}(\vec{q})}{\lambda_L{}^2 q^2 + K(\vec{q})} \quad , \tag{4}$$

where $\vec{H}(\vec{q})$, $\vec{M}(\vec{q})$, and $K(\vec{q})$ are the Fourier transforms of $\vec{H}(\vec{r})$, $\vec{M}(\vec{r})$, and $K(\vec{r})$, respectively, and $\vec{H}$ and $\vec{M}$ are assumed to be transverse modes. The molecular field acting on the magnetic moment at $\vec{r}$ is expressed as

$$\begin{aligned} \vec{H}_m(\vec{r}) &= \int d\vec{r}\,'\ \gamma(\vec{r} - \vec{r}\,')\ \vec{M}(\vec{r}\,') + \vec{H}(\vec{r}) \\ &= \int d\vec{r}\,'\ \tilde{\gamma}(\vec{r} - \vec{r}\,')\ \vec{M}(\vec{r}\,') \ . \end{aligned} \tag{5}$$

Here $\gamma(\vec{r} - \vec{r}\,')$ represents the coupling between the magnetic moments at $\vec{r}$ and $\vec{r}\,'$ in the normal state, while $\tilde{\gamma}(\vec{r} - \vec{r}\,')$ represents that in the superconducting state, including the screening effect by the persistent current. From (4) and (5) we obtain the relation between the Fourier transforms of γ and $\tilde{\gamma}$ as follows:

$$\tilde{\gamma}_{\vec{q}} = \gamma_{\vec{q}} - \frac{4\pi K(\vec{q})}{\lambda_L{}^2 q^2 + K(\vec{q})} \ . \tag{6}$$

Therefore, even when $\gamma_{\vec{q}}$ is ferromagnetic as described by

$$\gamma_{\vec{q}} = (T_m - Dq^2)/C \quad , \tag{7}$$

with a Curie constant C, a Curie temperature T_m and a stiffness constant D, the interaction $\tilde{\gamma}_{\vec{q}}$ in the superconducting state becomes strongest at the finite wave number

$$Q \sim (4\pi C/D)^{1/4}\ \lambda_L{}^{-1/2} \ , \tag{8}$$

because the long range interaction is screened by the persistent current. In this way, the spin-sinusoidal order coexisting with superconductivity occurs with the wave number Q at the critical temperature

$$T_Q = C\ \tilde{\gamma}_Q \ . \tag{9}$$

3. Surface of the Semi-Infinite System.

It can be shown that when a spin is located within λ_L from the surface, a persistent current cannot be sufficiently induced around the spin, so that the screening effect by the persistent current is weakened. Let us consider a semi-infinite ferromagnetic superconductor occupying the space of $x \geq 0$ (with the surface at $x = 0$). All the physical quantities are assumed to be uniform in the yz plane, and all the field vectors are limited to the transverse modes. The persistent current is now expressed as

$$\vec{J}(x) = -\frac{c}{4\pi\lambda_L^2}\int_0^\infty dx' \, [K(x - x') + R_s \, K(x + x')] \, \vec{A}(x') \, , \tag{10}$$

where R_s represents the reflection of the current by the surface, and $R_s = 1$ and 0 correspond to specular and diffuse reflections, respectively. Similarly, the effect of the surface on the molecular field in the normal state is expressed as

$$\vec{H}_m^0(x) = \int_0^\infty dx' \, [\gamma(x - x') + R_m \, \gamma(x + x')] \, \vec{M}(x') \, . \tag{11}$$

For $R_m > 0$, the coupling between magnetic moments at x and x' are enhanced with respect to that of the bulk system $\gamma(x-x')$, whereas for $R_m < 0$ it is suppressed. In the superconducting state, the molecular field is changed to

$$\vec{H}_m(x) = \vec{H}_m^0(x) + \vec{H}(x) = \int_0^\infty dx' \, \tilde{\gamma}_{eff}(x, x') \, \vec{M}(x') \, , \tag{12}$$

where $\vec{H}(x)$ is the magnetic field associated with $\vec{J}(x)$, and $\tilde{\gamma}_{eff}$ represents the effective coupling between magnetic moments. By assuming $R_s = 1$, $\tilde{\gamma}_{eff}$ is obtained from (2), (3), (10), (11), and (12) in the form [5]

$$\tilde{\gamma}_{eff}(x, x') = \tilde{\gamma}(x - x') - \tilde{\gamma}(x + x') + (1 + R_m) \, \gamma(x + x') \, . \tag{13}$$

It is found from (13) that in the limit of $x \gg \lambda_L$ and $x' \gg \lambda_L$, $\tilde{\gamma}_{eff}(x, x')$ tends to the screened interaction $\tilde{\gamma}(x - x')$ in the bulk system. For $x \sim \lambda_L$ and $x' \sim \lambda_L$, however, the screening effect by the persistent current is weakened, because the surface prevents the persistent current from being sufficiently induced around each spin. In the limit of $x \ll \lambda_L$ and $x' \ll \lambda_L$, $\tilde{\gamma}_{eff}$ of (13) is found to reduce to the unscreened interaction $\gamma(x - x') + R_m \, \gamma(x + x')$.

By means of a molecular-field approximation, we have studied the spontaneous spin order in the neighborhood of the surface. The spontaneous magnetization M(x) is obtained by solving

$$M(x) = N \, g \, \mu_B \, J \, B_J\left[\frac{g\mu_B J}{k_B T}\int_0^\infty dx' \, \tilde{\gamma}_{eff}(x, x') \, M(x')\right] \, , \tag{14}$$

where N is the number of the rare-earth ions per unit volume, g the Lande factor, μ_B the Bohr magneton, J the total angular momentum of the rare-earth ions, and $B_J[X]$ is the Brillouin function. For simplicity, we assume that the persistent current is local, so that we put $K(q) = 1$. We take, throughout this paper, the following values of the parameters: $\bar{c} = 4\pi C/T_m = 2.0$ and $d = D/\lambda_L^2 T_m = 0.01$. The critical temperature T_s of the surface spin order is obtained from (14) as $T_s = 0.87 \, T_m$ for $R_m = 1$ and $T_s = 0.76 \, T_m$ for $R_m = 0$ [6]. These values of T_s are higher than the critical temperature T_Q ($T_Q = 0.74 \, T_m$ with the present parameter values) of the spin-sinusoidal order in the bulk system, because of the decrease of the screening effect near the surface. The spatial variation of the spontaneous magnetization in the case of $R_m = 1$ is plotted in Figure 2 for various temperature values. For $T_s > T > T_Q$, the magnetization M(x) is bound mainly in the region

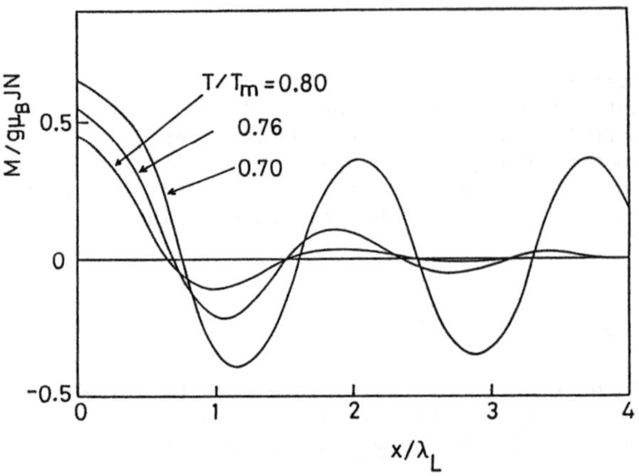

Figure 2. Spatial variation of the spontaneous surface magnetization.

$0 \leq x \leq \lambda_L$, and decays with oscillation when x increases further (see the curves for $T/T_m = 0.80$ and 0.76). For $T < T_Q$, on the other hand, the surface magnetization is connected smoothly with the spin-sinusoidal order in the bulk system (see the curve for $T/T_m = 0.7$).

Another unusual property of the surface is that when an external magnetic field H_0 is applied parallel to the surface above the critical temperature T_s (but below a certain temperature T_o, as shown below), the magnetic field penetrates from the surface with oscillation, as shown in Figure 1. By using the spin-spin interaction (13), and by assuming $R_m = 1$, the penetrating magnetic field H(x) is calculated, within a linear response to H_0, to be of the form [7]

$$H(x) = \frac{H_0}{q_2(1+q_2^2)-q_1(1+q_1^2)} [q_2(1+q_2^2)e^{iq_1x/\lambda_L} - q_1(1+q_1^2)e^{iq_2x/\lambda_L}] , \qquad (15)$$

where q_1 and q_2 are

$$q_1 = i\gamma_1, \quad q_2 = i\gamma_2 \quad , \quad \text{for } T > T_o ,$$

$$q_1 = k_1 + ik_2, \quad q_2 = -k_1+ik_2 \quad , \quad \text{for } T < T_o , \quad \text{with} \qquad (16)$$

$$\binom{\gamma_1}{\gamma_2} = [\ (1/2d) \ ((T/T_m) + d - 1 \mp [\{(T/T_m) - d - 1\}^2 - 4d\bar{c}]^{1/2}) \ ]^{1/2} , \qquad (17)$$

$$\binom{k_1}{k_2} = [(\frac{(T/T_m)+\bar{c}-1}{4d})^{1/2} \mp \frac{(T/T_m)+d-1}{4d} \ ]^{1/2} , \qquad (18)$$

$$T_o/T_m = 1 + d + 2 \ [d\bar{c}]^{1/2} \ . \qquad (19)$$

Therefore, H(x) decays monotonically with x for $T > T_o = 1.29\ T_m$, whereas it decays oscillatorily for $T < T_o$. It is also to be noted that $H(x)/H_0$ diverges at T_s, corresponding to the onset of the surface spin ordering.

Two remarks are worth making. In the first place, the present authors have recently shown that when a single vortex is introduced at x = R, the interaction between the vortex and the penetrating external field is proportional to H(R), so that it can be attractive for temperatures $T_o > T > T_s$ [8]. As a result, it may be possible that the vortices bound near the surface occur for H_0 below the lower critical field H_{c1} of the bulk system.

Secondly, when a vortex is introduced in the bulk system, the magnetic field $H_v(r)$ around the vortex also decays oscillatorily for $T_o > T > T_Q$ [2], similarly to

H(x) penetrating from the surface. It can be shown that the interaction between two vortices separated by the distance R is proportional to $H_v(R)$. Therefore, the inter-vortex interaction is always repulsive for $T > T_o$, and the mixed state occurs at H_{c1} through the second-order phase transition (denoted by type II/2 transition). On the other hand, the inter-vortex interaction can be attractive for $T < T_o$, so that a finite density of vortices are introduced discontinuously at H_{c1} through a first-order phase transition, which corresponds to type II/1 or type I transition.

4. Thin Film.

In this section we study the property of a film of ferromagnetic superconductor (occupying the region $0 \leq x \leq a$). When the thickness a is of the order of λ_L, the screening effect by the persistent current is weakened everywhere in the film. By method similar to that used in the preceding section, we can obtain the effective interaction $\tilde{\gamma}_{film}(x,x')$ between the magnetic moments in the film, and calculate the magnetization M(x) of the spontaneous spin order from molecular-field equation similar to (14). The critical temperature T_s' of the spontaneous spin order is shown in Figure 3 as a function of a/λ_L with R_m as a parameter [9]. It is found that T_s' depends strongly both on a and R_m. It is interesting to note that the pattern of the spontaneous spin order changes with the film thickness, as shown schematically in Figure 3. A spin pattern symmetric or antisymmetric with respect to the center of the film occurs, depending on the value of a: with increasing a the number of oscillations of the pattern increases.

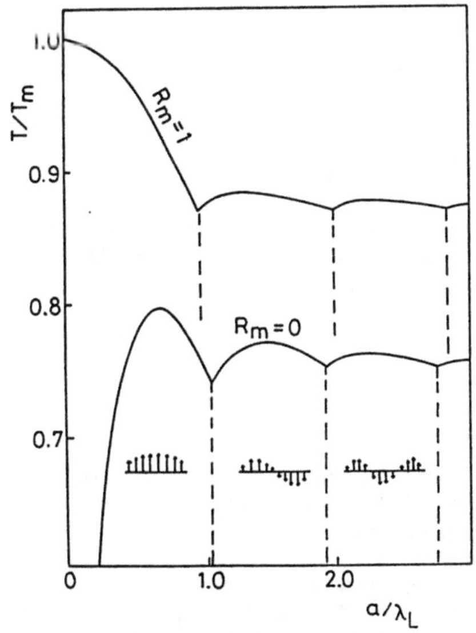

Figure 3. Critical temperature and ordering pattern of the spontaneous spin order in films.

When an external magnetic field H_0 is applied at the surfaces x = 0 and a, the magnetic field H(x) in the film can also be obtained [10], and the typical behavior of H(x) is shown schematically in Figure 1. For $T > T_o$, $H(x)/H_0$ is always positive, while for $T < T_o$ it can be negative, except for the case of very thin films.

In the following, we study the magnetization process of the film, especially in the vicinity of the lower critical field [11]. It is assumed that in the mixed state of the film the vortices parallel to the z axis (parallel to $\vec{H}_0$) are located at the position $\vec{R}_j$ in the xy plane. Then the persistent current is expressed as

$$\vec{J}(\vec{r}) = -\frac{c}{4\pi\lambda_L^2}\int d\vec{r}'\, K_{film}(\vec{r},\vec{r}')\,[\vec{A}(\vec{r}') - \frac{c\hbar}{e}\vec{\nabla}f(\vec{r}')] , \quad (20)$$

where the kernel $K_{film}(\vec{r},\vec{r}')$ can be obtained by assuming a reflection coefficient R_s of the current at the film surfaces, and $f(\vec{r})$ is half the phase of the superconducting order parameter. The function $f(\vec{r})$ satisfies the equation

$$\frac{\hbar c}{e}\vec{\nabla}\times\vec{\nabla}f(\vec{r}) = \phi\,\vec{e}_z\sum_j \delta^{(2)}(\vec{r}-\vec{R}_j) , \quad (21)$$

where ϕ is the unit flux and $\vec{e}_z$ is the unit vector in the z direction. The field quantities in the film are obtained by the solution of (2), (3), (20), (21), and a molecular-field equation similar to (14), where the argument of the Brillouin function is now replaced by $(g\mu_B J/k_B T)H_m(\vec{r})$ with the molecular field

$$H_m(\vec{r}) = H(\vec{r}) + \sum_{n=-\infty}^{\infty}\int d\vec{r}'[R_m^{|2n+1|}\gamma(x+x'+2na, y-y') + R_m^{|2n|}\gamma(x-x'+2na, y-y')]\,M(\vec{r}') . \quad (22)$$

In solving these equations, we divide $H(\vec{r})$ and $M(\vec{r})$ into their average values and the deviation: $H(\vec{r}) = \bar{H} + h(\vec{r})$ and $M(\vec{r}) = \bar{M} + m(\vec{r})$. Then we solve the values $\bar{H}$ and $\bar{M}$ exactly, and $h(\vec{r})$ and $m(\vec{r})$ in the linear approximation.

By using the field quantities thus obtained, we next calculate the Gibbs free energy of the system

$$G = E_s + W_{core} + \frac{\phi}{8\pi}\int d\vec{r}\, n(\vec{r})H(\vec{r}) + \frac{H_0}{8\pi}\int d\vec{r}[B(\vec{r}) - \phi n(\vec{r})]$$
$$+ \int d\vec{r}\,\{\frac{1}{2}M(\vec{r})H_m(\vec{r}) - k_B TN\log Z_J[\frac{g\mu_B J}{k_B T}|H_m(\vec{r})|]\} - \frac{H_0}{4\pi}\int d\vec{r}B(\vec{r}) , \quad (23)$$

where E_s is the superconducting condensation energy, W_{core} is the core energy of vortices, $B(\vec{r}) = H(\vec{r}) + 4\pi M(\vec{r})$ is the magnetic induction, and n and Z_J are defined by

$$n(\vec{r}) = \sum_j \delta^{(2)}(\vec{r}-\vec{R}_j) , \quad (24)$$

$$Z_J[X] = \sinh(\frac{2J+1}{2J}X) / \sinh(X/2J) . \quad (25)$$

The free energy G is minimized with respect to changes in the configuration of vortices. Then we can calculate the averaged value of magnetization as a function of the external field H_0, so that we obtain the magnetization curve.

In Figure 4 we show an example of the magnetization curves calculated for various values of the film thickness with the temperature fixed at $T/T_m = 1.0$. In this calculation, we take account of the effect of nonlocal persistent current by using $R_s = 1$ and the Ginzburg-Landau parameter $\kappa = 2.8$; we also assume $Ng\mu_B J\lambda_L^2/\phi = 0.3$ and $R_m = 1$. In Figure 4 the magnetization curve of the bulk system (dashed curve) is also plotted for the sake of comparison. From this figure, we can see several unusual features of the magnetization curve of films. In the first place, the Meissner effect cannot be found in thin films, although it occurs in the bulk system below the lower critical field H_{c1}. This is due to the effect of the external magnetic field penetrating from the film surfaces. The penetrating field not only makes the Meissner effect imperfect but also induces the paramagnetic magnetization of the rare-earth spins. Therefore, the magnetization curve of thin films shows only paramagnetic magnetization, even below the lower critical field.

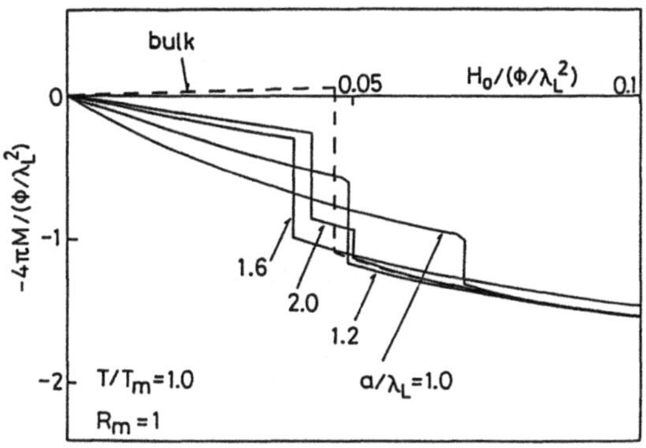

Figure 4. Magnetization curves of films for various values of the thickness.

Secondly, the lower critical field of films depends strongly on the film thickness, and in some cases it can be lower than that of the bulk system. This is unusual because in nonmagnetic superconductors the lower critical field of films is always higher than that of the bulk system. In nonmagnetic superconductors, the penetrating magnetic field $H(x)/H_0$ has always a positive sign, but in ferromagnetic superconductors it can be negative, as already discussed. Then the interaction between the penetrating field and the vortex becomes attractive, very similarly to the case of the semi-infinite system. Therefore the vortices in the film can be more stable than those in the bulk system, and it is possible to have values of H_{c1} lower than those of the bulk crystal.

Thirdly, some discontinuities occur in the magnetization curve of films in addition to the discontinuity at H_{c1}. This additional discontinuity originates from the discontinuous change of the vortex-lattice structures. In the case of $a/\lambda_L = 2.0$, the mixed state with a single vortex array at the film center occurs just at H_{c1}, but with increasing external field it changes discontinuously to the mixed state with double-vortex arrays.

Finally, the type of the phase transition at H_{c1} depends on the film thickness. In Figure 4 it is found that for $a/\lambda_L = 2.0$ and 1.6 the mixed state occurs discontinuously at H_{c1} through a type II/1 transition. However, with decreasing film thickness (see the cases of $a/\lambda_L = 1.2$ and 1.0) the mixed state occurs continuously at first, but immediately after that a discontinuity occurs. Such a transition is classified as a type II/3 transition.

In the case of type II/1 transition, the interaction between vortices has an attractive component. In the type II/3 transition, on the other hand, the inter-vortex interaction becomes always repulsive, but the strength of the repulsive interaction changes oscillatorily with the inter-vortex distance, partly because the saturation of the magnetization is due to the penetrating external field and partly because the screening effect by the persistent current is weakened. Then the mixed state occurs continuously, but with increasing external magnetic field the inter-vortex distance changes discontinuously because of the potential barrier between the two inter-vortex distances. We predict theoretically that in thin films with thickness of the order of λ_L a type II/3 transition will be observed, although it cannot be observed in bulk systems. More detailed investigation on the type II/3 transition, including the check of the validity of our approximate calculation, is now in progress [11].

5. Concluding Remarks

We have predicted theoretically many unusual properties of surfaces and films of ferromagnetic superconductors. In this paper, we have confined ourselves to the static properties. Some dynamical properties were also predicted by the present authors and published elsewhere [12]. Up to now only a limited number of experimental observations have been carried out for surfaces and films of ferromagnetic superconductors [13], but we hope that our theoretical predictions will be confirmed experimentally in the future.

References

1. See, for instance, Superconductivity in Magnetic and Exotic Materials, ed. by T. Matsubara and A. Kotani (Springer, Berlin, Heidelberg, New York, 1984).
2. A. Kotani and K. Okada, in Superconductivity in Magnetic and Exotic Materials, ed. by T. Matsubara and A. Kotani (Springer, Berlin, Heidelberg, New York, 1984), p. 64.
3. H. Matsumoto, H. Umezawa, and M. Tachiki, Solid State Commun. 31, 157 (1979); M. Tachiki, A. Kotani, H. Matsumoto, and H. Umezawa, Solid State Commun. 31, 927 (1979).
4. E. I. Blount and C. M. Varma, Phys. Rev. Lett. 42, 1079 (1979).
5. A. Kotani, S. Takahashi, T. Koyama, M. Tachiki, H. Matsumoto, and H. Umezawa, J. Phys. Soc. Jpn. 50, 3254 (1981).
6. T. Koyama, A. Kotani, S. Takahashi, M. Tachiki, H. Matsumoto, and H. Umezawa, J. Phys. Soc. Jpn. 51, 3469 (1982).
7. A. Kotani, M. Tachiki, H. Matsumoto, H. Umezawa, and S. Takahashi, Phys. Rev. B 23, 5960 (1981).
8. K. Okada and A. Kotani, J. Phys. Soc. Jpn. 54, 1944 (1985).
9. S. Takahashi, A. Kotani, M. Tachiki, H. Matsumoto, and H. Umezawa, J. Phys. Soc. Jpn. 52, 989 (1983).
10. A. Kotani, S. Takahashi, M. Tachiki, H. Matsumoto, and H. Umezawa, Solid State Commun. 37, 619 (1981).
11. K. Okada and A. Kotani, to be published.
12. K. Okada and A. Kotani, Solid State Commun. 46, 811 (1983); Solid State Commun. 54, 167 (1985).
13. For instance, T. Fujita, S. Ikegawa, and T. Ohtsuka, Proceedings LT-17 (Elsevier Science, Amsterdam, 1984), p. 335.

Part II

Surface-Sensitive Experimental Techniques

On the Theory of Electron Scattering and Emission from Magnetic Surfaces

R. Feder

Department of Physics, Rice University, Houston, TX 77251, USA
(on leave from Universität Duisburg, Fed. Rep. of Germany)

The current status of theoretical approaches to various electron scattering and emission mechanisms from magnetic surfaces is outlined. For SPLEED and photoemission numerical results are presented to illustrate the role of theory in extracting quantitative information on surface magnetism from experimental data.

1. Introduction.

Recent experimental progress [1] in studying magnetic properties of surfaces and of thin films by a variety of electron spectroscopy techniques (see in particular the articles by ALVARADO et al. [2], GRADMANN [3], PIERCE [4] and RAU [5] and Chapters 5 to 13 of Reference 1) has been paralleled by theoretical advances of two types: (a) first-principles calculations of the electronic and magnetic structure of the ground state [6-9] and (b) realistic treatment of the electron-surface interaction and quantitative calculations of spin- dependent electron scattering and emission cross sections [10].

The latter type of theory, which involves excited states, is not only interesting per se but also of importance for experiment: first because it extracts detailed information on magnetic properties from experimental data, and second because it provides guidance for future experimentation. In this article, the current status of theoretical approaches is outlined and illustrative numerical results are presented and discussed in conjunction with corresponding experimental data. Section 2 addresses the general framework of an effective one-electron picture. Sections 3 and 4 are devoted to electron scattering (in particular elastic spin-polarized low-energy electron diffraction, SPLEED) and photoemission (and its inverse, bremsstrahlung), respectively. Section 5 summarizes the current status of theory and computational implementation, with a view to future developments.

2. Effective One-Electron Picture.

The various electron spectroscopic methods can be described as scattering problems with an electron in at least one of the input and output channels. The interaction of this electron with the magnetic surface system constitutes a complicated many-body problem, which can -- on the formal basis of excited-state density functional theory -- be reduced to an effective one-electron problem involving as an "optical potential" a complex, energy-dependent and non-local self-energy matrix, which is a functional of the charge and spin density of the ground state. (For details of this and the following, see the article by FEDER [10] and references therein.) For numerical tractability by present-day means, one has to resort to a local approximation, i.e. the interaction is described by a complex electrostatic potential $V(E,\vec{r})$ and effective magnetic field $\vec{B}(E,\vec{r})$, with the latter taking account of the magnetic exchange interaction. To incorporate spin-orbit coupling, which produces observable effects even for low-Z materials (like Fe, with Z = 26), on an equal footing, it is most expedient to employ a Dirac-type equation

$$[H_D^o + V(E,\vec{r}) - \beta\, \vec{\sigma}\cdot\vec{B}(E,\vec{r})]\ \psi(\vec{r}) = E\psi(\vec{r}) \ , \qquad (1)$$

where H_D^o is the usual free-electron Dirac Hamiltonian, $\beta = \bar{\sigma}_3 \times 1_2$, $\vec{\bar{\sigma}} = (\bar{\sigma}_1, \bar{\sigma}_2, \bar{\sigma}_3)$, $\bar{\sigma}_k = 1_2 \times \sigma_k$, and σ_k are the (2x2) Pauli spin matrices. If $\vec{B}$ is parallel to a global quantization axis and has the same (position-dependent) values in all unit cells, this corresponds to a band (Stoner) model of ferromagnetism, which is adequate at low temperatures. The modern "fluctuating-local-moment" picture of transition-metal ferromagnetism entails $\vec{B}$ varying from cell to cell both in magnitude and direction, with some short-range correlation. The interaction in (1) becomes more transparent if one goes to the non-relativistic limit and neglects spin-orbit coupling: one then has two decoupled Schrödinger equations, one with a "spin-up" potential V^+ and one with a "spin-down" V^- defined as $V^{\pm} = V \mp B$.

The construction of effective potentials, which are adequate for realistic calculations, is rather complicated because of exchange and correlation. Referring for details to the article by FEDER [10], it may suffice here to say that for the real parts systematic local-density approximations have been developed involving an exchange contribution, which decreases with increasing energy; for the imaginary parts, which account for finite lifetimes, viable first-principles procedures do not yet exist (except at $E = E_F$, where $V_{im} = 0$), and one has to resort to expressions with adjustable parameters (e.g. for energies between 40 and 200 eV above the vacuum zero, V_{im} takes, universally, a value of the order of 4 eV). A schematic picture of the real part of V, which is dominated by the electrostatic contribution is shown in Figure 1, with Coulomb singularities at the nuclear sites and a "surface potential barrier" going (with image asymptotics z^{-1}) to zero on the vacuum side.

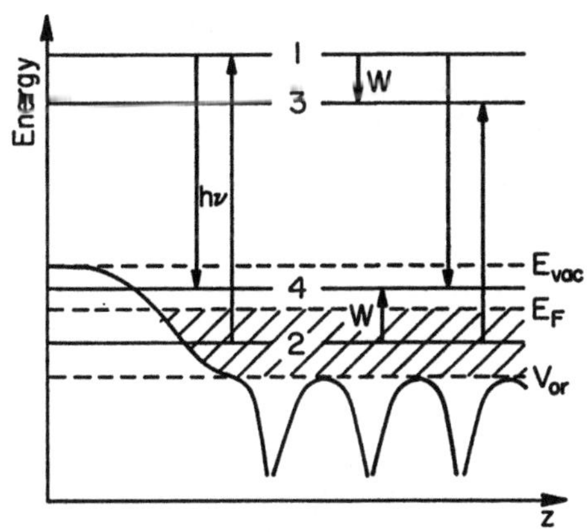

Figure 1. Schematic picture of the real part of the effective one-electron potential along a line z normal to the surface and energy levels relevant for electron scattering and photoemission (see text). (The hatched area below the Fermi energy E_F indicates the occupied states.)

Numerical solution of (1) with appropriate boundary conditions provides "half-space" one-electron states (e.g. the four energy levels labelled 1-4 in Figure 1). By means of these, the cross-sections for the various spectroscopic methods can be obtained, as will be indicated in the following sections.

3. Spin-Polarized Low-Energy Electron Scattering.

3.1 Elastic Scattering.

The theoretically simplest electron spectroscopy is elastic scattering of an electron beam of well-defined energy E, well-defined momentum parallel to the surface (or angles of incidence), and well-defined spin polarization. For targets with perfect two-dimensional lattice periodicity parallel to the surface, this is (elastic) SPLEED. The one half-space state (e.g. 1 in Figure 1), which is relevant, can be numerically calculated by means of relativistic multiple scattering theory [10] and provides the intensities of the diffracted beams as functions of energy E, polar and azimuthal angles of incidence ν and ϕ, spin polarization of the incident beam and layer-dependent magnetization M_n (with n = 1 referring to the topmost monatomic layer parallel to the surface). The most convenient observable quantity for studying magnetic structure is an exchange-induced scattering asymmetry A_{ex}, which in the absence of spin-orbit coupling is $A_{ex} = (I^+-I^-)/(I^++I^-)$, where $I^+(I^-)$ is the intensity of the diffracted beam under consideration for the spin of the incident beam parallel (antiparallel) to the majority spins of the target. (For the definition in the realistic case with spin-orbit coupling, see Reference 10). Clearly, the actual set $\{M_n(T)\}$ is "coded" in $A_{ex}(E, \nu, \phi; T)$ at any given temperature T. Its extraction from measured A_{ex} profiles is, however, in general not possible directly, but requires detailed comparison with theoretical A_{ex} features, which are calculated for assumed trial sets $\{M_n\}$ and respond sensitively to changes in M_n. This procedure yields $\{M_n(T)\}$ for any experimentally chosen temperature. Extrapolation to $\{M_n(T=0)\}$, the quantities obtained by ground-state theory [6] is possible by means of an effective Heisenberg model [11]. As T approaches the Curie temperature T_c, the excess of the magnetic coherence length over the electron probing depth (typically about 10 layers) renders A_{ex}, in an increasingly good approximation, proportional to M_1 [12] and thus permits a direct experimental determination of surface critical exponents [2].

The determination of M_1 at lower temperature is illustrated in Figure 2 for Ni(001). The calculated A_{ex} "rocking curve" of the (10) beam at E = 69 eV is seen to be very sensitive to the assumed $\{M_n\}$. Comparison with experiment suggests preference for a set $M_n(T)$, which at T = 0 has M_1 slightly enhanced over the bulk value. Obviously, such magnetic structure analysis should be based on more than one A_{ex} profile. To make the T = 0 extrapolation more reliable, $\{M_n(T)\}$ should be determined for several T if the lowest experimentally accessible T is already a sizeable fraction of T_c. For Fe(110), a preliminary SPLEED analysis suggests [3] an enhancement of M_1 by about 30% over M_{bulk}. Calculations [13] for Fe(110)-S indicate the power of SPLEED for studying the influence of chemisorption on surface magnetism, and make corresponding measurements highly desirable.

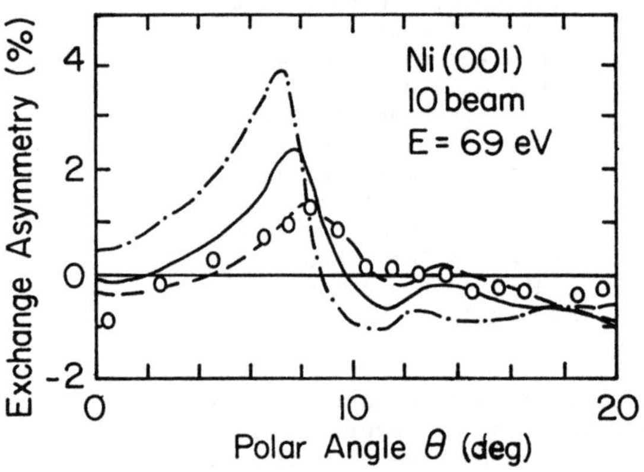

Figure 2. SPLEED exchange asymmetry $A_{ex}(\nu)$ of the (10) beam from Ni(001) and T = 300 K: experiment (ooo) and theory for magnetization models which at T = 0 have $M_1 = 1.05\ M_b$ (bulk) (---), $M_1 = 0.80\ M_b$ (___), and $M_1 = 0$ (dead layer) (-·-·-). (From Reference 11).

3.2 Inelastic Scattering.

The theoretical treatment obviously requires at least two quasi-electron half-space states: one with ingoing boundary condition, differing in energy by an energy loss w from one with outgoing boundary condition (e.g. states labelled 1 and 3 in Figure 1). Further ingredients depend on the nature of the loss mechanism. For magnetic systems, spin-dependent mechanisms are of special importance: excitation of electron-hole pairs and excitation of (collective) spin fluctuations (in particular magnons).

The former mechanism involves two further quasi-electron states (2 and 4 in Figure 1). To first order in an effective Coulomb interaction V, the scattering amplitude has a direct part $<34|V|12>$ [cf. Figure 3(a)] and an exchange part $<43|V|12>$ [cf. Figure 3(b)]. If spin-orbit coupling is neglected, the four states

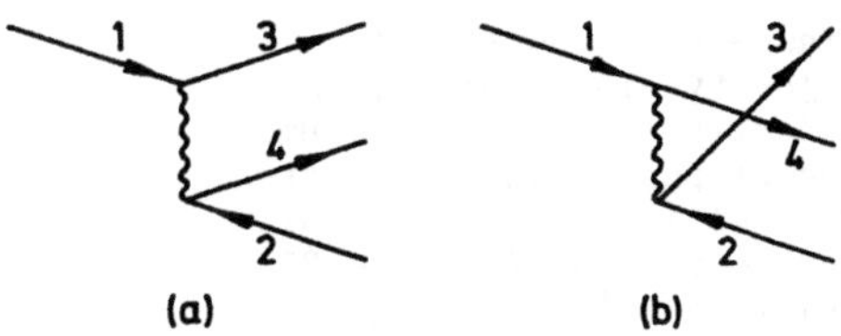

Figure 3. Electron-hole excitation by (a) direct and (b) exchange scattering. The wiggly line represents an effective Coulomb interaction.

have definite "spin-up" or "spin-down" (spin labels σ_1 = +/- etc.), and if furthermore V has no spin-mixing part, spin is conserved at each vertex. The direct process is then always a "non-flip" process ($\sigma_3 = \sigma_1$), while the exchange process permits "spin-flip" ($\sigma_3 = -\sigma_1$) and creation of an opposite-spin electron-hole pair ($\sigma_4 = -\sigma_2$) ("Stoner excitation"). Calculation of the above matrix elements using realistic half-space states is rather demanding and has not yet been accomplished. Experimentally, "flip" and "non-flip" scattering rates have recently been obtained for 3d ferromagnets by using polarized incident electrons and analyzing the spin of the outgoing electrons [14]. Calculation of these quantities is further complicated by the fact that they involve a summation over pairs (2,4).

For magnon excitation, a theoretical approach has been formulated by MILLS [15], but realistic calculations have not yet been performed. (For some discussion and reference on this subject see Reference 10). With the exciting prospect of surface magnon dispersions coming within experimental reach, further theoretical efforts appear very timely.

4. Photoemission and Bremsstrahlung.

As schematically indicated by the two vertical arrows in the left-hand part of Figure 1, two quasi-particle states are involved: 1 and 2 in the case of photoemission, and 1 and 4 in the case of bremsstrahlung (inverse photoemission). A theoretical treatment can be based on the "golden-rule" expression $\langle 1|H'|2\rangle$ (or $\langle 1|H'|4\rangle$), where $H' = \vec{A}\cdot\vec{p}$ (nonrelativistic) or $H' = \vec{A}\cdot\vec{\alpha}$ (relativistic) is the interaction with the electromagnetic radiation field (characterized by the magnetic vector potential $\vec{A}$). A somewhat more general approach, which incorporates a priori the finite lifetime of the lower state (2 or 4), is of a scattering-theory nature [10]. Schrödinger- and Dirac-equation-based versions of the theory have been formulated, computationally implemented, and successfully applied to calculating spin-, angle-, and energy-resolved photoemission intensities. For the 3d ferromagnets it appears that relative to the present level of experimentation (in particular energy and angular resolution) the less demanding nonrelativistic version (which neglects spin-orbit coupling) is adequate except for describing some special features.

The state of the art of photoemission calculations and their power in extracting information from experimental data is illustrated by a recent study on the Ni(110)(2x1)-O chemisorption system [16]. The geometry is characterized by a "saw-tooth" reconstruction of the two topmost Ni layers (with alternate [001] rows missing in these two layers) and O absorbed in long-bridge position along [001]. Spin-resolved normal photoemission measurements (at T = 300K) [17] show the following changes upon O-chemisorption: (a) an overall intensity reduction, which is stronger for minority than for majority spin electrons and (b) a drastic reduction of the separation of majority and minority peaks ("exchange splitting"). Theory not only correctly reproduces these changes, but provides a deeper understanding by exploring the "evolution" of the spectra via hypothetical "intermediate" models (see Figure 4). While (1x1) surface relaxation has almost no effect, the clean (2x1) reconstruction selectively reduces the dominant S_4-symmetry peaks. Subsequent O adsorption selectively reduces the S_3-symmetry peaks. These changes -- coming out of the full-fledged one-step theory photoemission calculation -- are qualitatively plausible on the grounds of simple tight-binding arguments. With regard to the layer-dependent magnetization, rather striking results were obtained: (1) quenching the magnetization of the two top Ni layers

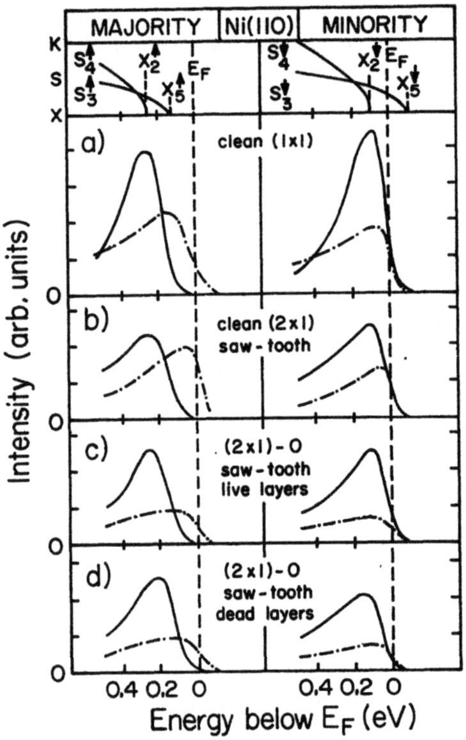

Figure 4. Theoretical majority (left column) and minority spin (right column) normal photoemission spectra from Ni(110) due to normally incident light ($h\nu$ = 16.85 eV) with E || [$1\bar{1}0$] (——) and E || [001] (-·-·-) for the following surface models: (a) clean (1x1) with $M_1 = M_b$, (b) clean (2x1) saw-tooth reconstruction with $M_1 = M_b$, (c) (2x1)-O saw-tooth model with $M_1 = M_b$, and (d) as (c) but $M_1 = M_2 = M_3 = M_4 = 0$, i.e. four dead layers. Topmost panel: self-energy-corrected bulk band structure along the KX direction (from Reference 16).

has practically no effect, (2) the observed reduction of the exchange splitting is obtained by $M_3 = M_4 = 0$ [cf. Figure 4(d)]. The latter finding indicates at least, that the S_4 splitting in layers 3 and 4 is quenched. To actually determine M_3 and M_4, a SPLEED analysis appears desirable.

For a review of further theoretical photoemission and bremsstrahlung studies at room temperature and near the Curie temperature (determining the degree of short-range magnetic order in 3d ferromagnets), the reader should consult the article by CLAUBERG and FEDER [18].

5. Outlook.

For elastic electron scattering as well as for (valence-band) photoemission and bremsstrahlung (inverse photoemission), theory has attained a high level of sophistication. Computer codes exist to perform realistic calculations, which permit the extraction of quantitative information on magnetic surface properties from experimental data. For other interesting processes and the promising techniques based on them (e.g. spin-dependent inelastic scattering, secondary electron emission, Auger electron emission, spin-polarized metastable atom deexcitation [19], electron capture spectroscopy [5,19], and spin-polarized low-energy positron scattering [19]), the theoretical treatment is more complex and still at a prequantitative level. The time is, however, ripe for full-scale theoretical developments in these areas.

References

1. Polarized Electrons in Surface Physics, ed. by R. Feder (World Scientific, Singapore, 1985).
2. S. F. Alvarado, E. Kisker, and M. Campagna, this volume.
3. U. Gradmann, this volume.

4. D. T. Pierce, this volume.
5. C. Rau, this volume.
6. A. J. Freeman, C. L. Fu, S. Onishi, and M. Weinert, in Polarized Electrons in Surface Physics, ed. by R. Feder (World Scientific, Singapore, 1985), chapter 1.
7. G. Allan, this volume.
8. A. J. Freeman and C. L. Fu, this volume.
9. R. H. Victora, this volume.
10. R. Feder, in Polarized Electrons in Surface Physics, ed. by R. Feder (World Scientific, Singapore, 1985), chapter 4.
11. R. Feder, S. F. Alvarado, E. Tamura, and E. Kisker, Surface Sci. 127, 83 (1983).
12. R. Feder and H. Pleyer, Surface Sci. 117, 285 (1982).
13. E. Tamura and R. Feder, Surface Sci. 139, L191 (1984).
14. J. Kirschner, in Polarized Electrons in Surface Physics, ed. by R. Feder (World Scientific, Singapore, 1985), chapter 8.
15. D. L. Mills, J. Phys. Chem. Solids 28, 2245 (1967).
16. R. Feder and H. Hopster, Solid State Commun. 55, 1043 (1985).
17. W. Schmitt, H. Hopster, and G. Güntherodt, Phys. Rev. B 31, 4035 (1985).
18. R. Clauberg and R. Feder, in Polarized Electrons in Surface Physics, ed. by R. Feder (World Scientific, Singapore, 1985), chapter 14.
19. F. B. Dunning, C. Rau, and G. K. Walters, Comments on Solid State Physics, to be published.

Electron Scattering and Synchrotron Radiation in Surface Magnetism

S.F. Alvarado, E. Kisker, and M. Campagna

Institut für Festkörperforschung der KFA Jülich,
D-5170 Jülich, Fed. Rep. of Germany

Some of the fundamental questions of magnetism deal with the short- and long-range magnetic order of ferromagnets near the Curie temperature T_C. Of special current interest is the magnetism of ultrathin films and of the topmost atomic layers of macroscopic thick samples as well as the relationship between the electronic structure and the magnetic order of itinerant ferromagnets like Fe, Co, and Ni.

Spin-polarized spectroscopy is one of the experimental techniques which accomplishes to study the aforementioned phenomena and allows to test theoretical models most directly. In this paper we focus our discussion on experimental results obtained by means of spin-polarized electron diffraction and spin-polarized photoemission. The size of the samples is such that the magnetic coherence length perpendicular to the surface $\xi_{perp}(T) << d$ where d is the thickness of the specimen, i.e., we are dealing with semi-infinite systems. The experiments were performed using remanently magnetized crystals so that data collection could be done in the absence of an external magnetic field. The magnetization vector lay in the surface plane.

1. Probing Long-Range Magnetic Order by SPLEED.

In a SPLEED experiment the quantity related to the long-range magnetic order of the surface is the magnetic-exchange asymmetry A_{ex} of electrons diffracted on the crystal. This quantity can be determined, as done for the cases discussed here, by measuring the electron intensity diffracted along a fixed direction for antiparallel ($I_{\uparrow\uparrow}$) and parallel ($I_{\uparrow\downarrow}$) orientation of the electron spin-polarization of the incident electrons and the sample magnetization:

$$A_{ex} = (I_{\uparrow\uparrow} - I_{\uparrow\downarrow})/ (I_{\uparrow\uparrow} + I_{\uparrow\downarrow}).$$

The surface sensitivity is determined by the electron mean-free path, which is of the order of one lattice constant for kinetic energies in the range $10 \leq E \leq 200$ eV. Thus this technique, as well as photoemission, can probe the second and third atomic layers below the surface. Still the behavior intrinsic to the topmost atomic layer can be measured because the magnetic coherence length behaves as $\xi_{perp}(T) = \xi_{o,perp}(1-T/T_C)^{-\nu}$, where $\nu = 0.7$ [1] diverges near T_C, i.e., the layers in the region $d_s << \xi_{perp}(T)$ exhibit the same temperature dependence of the magnetization as the topmost atomic layer. Near T_C the asymmetry A_{ex} is expected to be proportional to the magnetization. This linear relationship has been experimentally tested by performing measurements at different electron energies and scattering angles, yielding the same temperature dependence of the surface layer magnetization near T_C. This experimental result has also been confirmed in a theoretical analysis by FEDER and PLEYER [2].

We shall interpret the experimental results in the frame of theoretical results derived from simple models with localized magnetic moments, being aware that matter is more complicated! The simplest model we may recur to is the simple-cubic semi-infinite Ising model illustrated at the left panel of Figure 1. The right panel shows the phase diagram predicted by mean-field theory. In the Figure, SP, SF, (BP,BF) refer to surface (bulk) being para- or ferromagnetic. The lines O, S, and E describe second-order transitions called respectively ordinary, surface, and

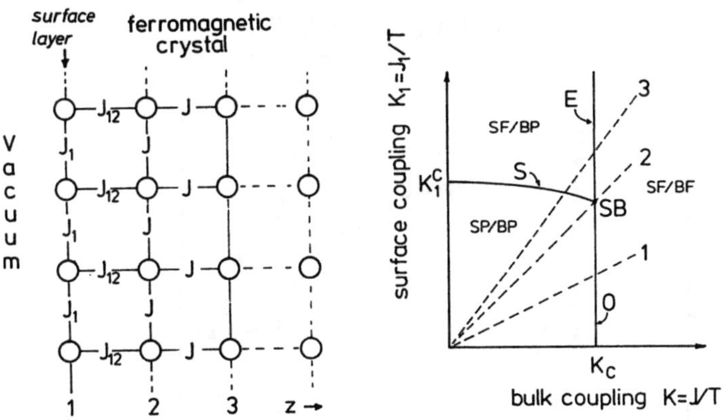

FIGURE 1.

extraordinary. The label SB indicates a multicritical point which corresponds to the special transition. The paths denoted 1, 2, or 3 represent temperature trajectories at constant $J_1/J < K^c{}_1/K_c$, $J_1/J = K_1/K_c$, or $J_1/J > K_1/K_c$ repectively. For systems exhibiting a continuous phase transition in the bulk, one can see from this diagram that for J_1 below a certain critical value J_{1c} the surface and the bulk share the same critical temperature ($T_{cs} = T_{cb}$, ordinary transition), while for $J_1 > J_{1c}$ surface magnetic order should occur in the presence of a paramagnetic bulk ($T_{cs} > T_{cb}$, surface transition). According to renormalized-field group results [3-5] and Monte Carlo calculations [6], the magnetization of the topmost atomic layer vanishes as $m_1 \propto t^{\beta_1}$, where $t = (1-T/T_c)$, with a critical exponent whose value is $\beta_1 \cong 0.78$ for the ordinary transition. For the surface transition one expects $\beta_1 = 1/8$ which is the value calculated for the 2D-Ising system [7]. For XY- or Heisenberg 2D-systems no long-range order is expected [8], and the surface transition should not occur (under normal conditions) in such semi-infinite systems either. In the ordinary transition, however, the value of β_1 is slightly larger than for the Ising case. Thus $\beta_1 \cong 0.82$ (XY) and $\beta_1 \cong 0.84$ (Heisenberg) [4-6,9].

Figures 2-4 show SPLEED results obtained with Ni single crystal surfaces [10,11] and on Gd(0001) epitaxial thin films [12]. In the case of Ni it is found that $T_{cs} = T_{cb}$ and $\beta_1 = 0.8 \pm 0.02$, which shows that the surface region undergoes an ordinary transition. In the case of Gd, however, the surface Curie point is found to lie up to 22 K above the bulk value T_{cb} = 293 K determined in situ by means of a magneto-optic Kerr effect (MOKE) technique. According to the phase diagram in Figure 1, a surface transition occurs. It should be noted that this result suggests the presence of an anisotropy at the surface which makes magnetic ordering

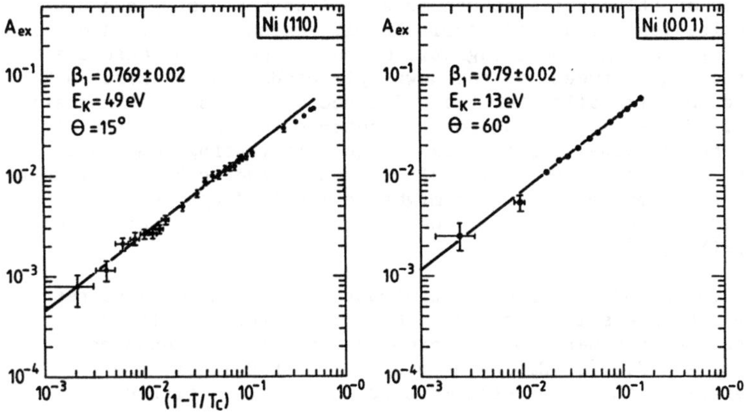

FIGURE 2.

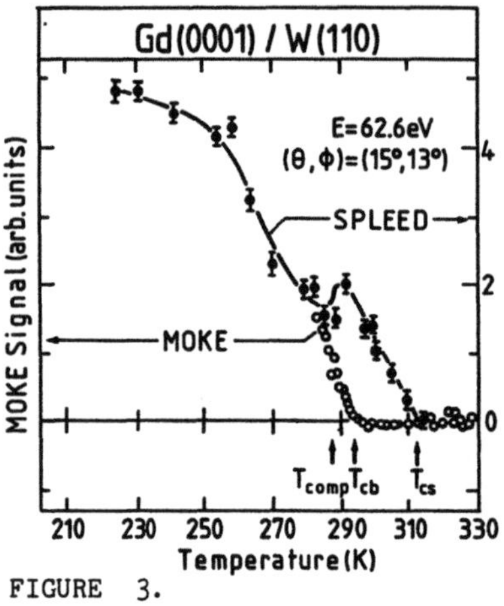

FIGURE 3.

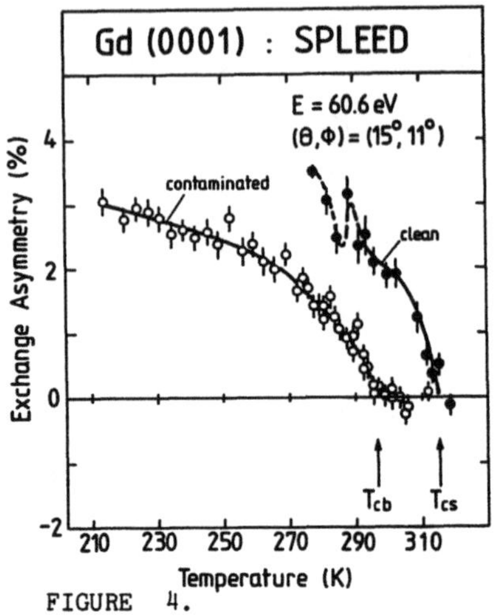

FIGURE 4.

possible above T_{cb}. If the surface were magnetically isotropic, as bulk Gd seems to be in the critical region, then no ordering above T_{cb} would be expected. Figure 4 illustrates the effect of surface contamination on the surface transition of Gd(0001). It is found that T_{cs} decreases continuously with coverage until the bulk critical point is reached. The same behavior is found by oxygen or hydrogen deposition on the surface. These results can be interpreted according to the phase diagram of Figure 1 as to be caused by an extrinsically induced reduction of the surface coupling J_1.

2. Short-Range Magnetic Order in Spin-Polarized Photoemission.

Spin-resolved spectroscopy has become an important new tool in the study of the electronic structure of magnetic materials [13-16]. In a photoemission experiment with angle-, energy-, and spin-resolution it is possible to determine the spin-split density of initial states, ($n_\uparrow$ and $n_\downarrow$), by measuring the photoelectron polarization $P(E_b) = (n_\uparrow - n_\downarrow)/(n_\uparrow + n_\downarrow)$ and the energy distribution curve (EDC) $N(E_b) = n_\uparrow + n_\downarrow$ as a function of the binding energy E_b. Furthermore, by performing experiments with synchrotron radiation, the energy dispersion of the spin-split states can also be measured. This technique has been recently applied to reach a deeper understanding of the relationship between the electronic structure and the ferromagnetism of 3d transition metals.

It is currently accepted that the 3d ferromagnets Fe, Co, and Ni are best described, at T = 0 K, by an itinerant-electron model. A controversial question is: How does the magnetization, at a microscopic level, vanish near T_C? According to the Stoner model the size of the local magnetic moment decreases because the exchange energy separation Δ_{ex} between up- and down-spin bands goes to zero as T approaches T_C. Improved theories allow for the existence of finite local magnetic moments at and above T_C. For Fe the discussion is centered currently on the size of the magnetically ordered domains. If the size of the fluctuating domains is large enough, one might refer to the so-called "local band structure" [17], whereas in the absence of short-range order the system might be modelled by the "disordered-local-moment" picture [18]. A synthesis between these extreme models has been proposed by MORIYA [19].

By spin-resolved photoemission the electronic structure of a ferromagnet near T_C can be probed since the exchange splitting can be observed directly. Theories predict that the temperature dependence of the exchange splitting depends on the initial-state wave-vector k [18,21]. Employing synchrotron radiation, the temperature-induced changes in the spin-resolved EDCs occurring at different

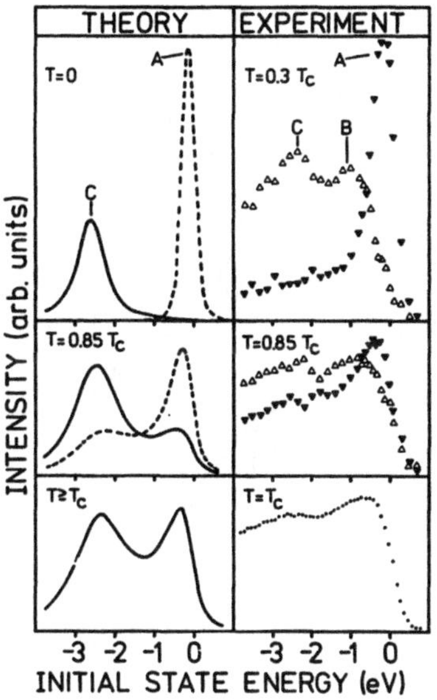

FIGURE 5.

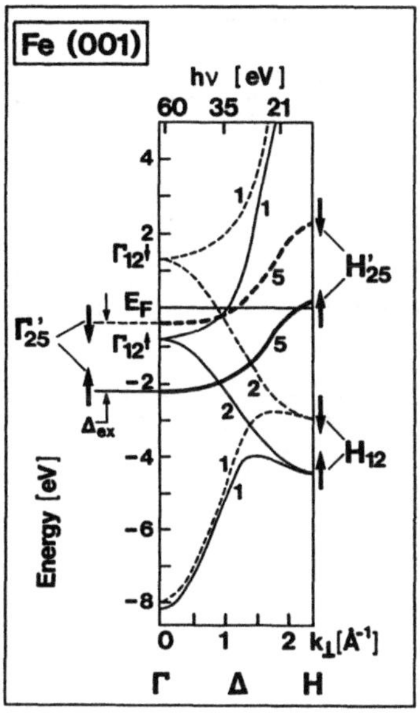

FIGURE 6.

k-vectors along the delta-line of the Brillouin zone have been recently determined [16]. Figure 5 (right panel) displays spin-resolved EDCs taken at 60 eV photon energy with initial states near Γ, taken at $T/T_C = 0.3$ and $T/T_C = 0.85$. At room temperature, emission from $\Gamma'^{\uparrow}_{25}$, $\Gamma'^{\downarrow}_{25}$, and $\Gamma^{\uparrow}_{12}$ can be identified [22]. The separation between the exchange split peaks amounts to 2.2 eV. Upon heating, characteristic changes occur in the EDCs: A structure emerges in the minority-spin EDC at 2.5 eV binding energy, while the peak intensity near E_F decreases dramatically.

Similar but much less pronounced changes occur in the majority-spin EDC. It has been shown that the occurrence of the new peaks in the spin-resolved EDC arise from a reversed magnetization in regions of an extension Λ, which is large enough to build up a well-defined local-band structure [23] and that the decrease in the intensity of the minority-spin peak near E_F is indicative of a short-range order length smaller than the coherence length of the photoelectron. The detailed analysis of data in a new cluster calculation by HAINES, CLAUBERG, and FEDER [24] allowed to determine that short-range order exists within regions with $\Lambda \geq 4$ Å. The left panel of Figure 5 shows results of their calculations which are to be compared with experimental data in the right panel. For smaller values of Λ the observed exchange-split peaks are expected to merge into a central peak near the nonmagnetic Γ'_{25} energy [24]. This "collapse" of the observed exchange energy of the Γ'_{25} bands indicates that most of the electrons have a group velocity high enough to average over the local exchange fields. Data taken for initial-state wave vectors in the right half of the Brillouin zone display a completely different behavior. Rather than a decrease of the minority-spin intensity near E_F, a strong increase is observed [16]. Since at this specific wave vector the Δ_5 band is above E_F, this might be indicative of a collapse of the exchange splitting at this wave vector in agreement with theoretical predictions [18,20]. For a detailed analysis of these data we refer to the literature [23,25]. The behavior of the minority-spin band above E_F can be tested most directly by the new technique of spin-polarized inverse photoelectron spectroscopy [26,27]. Indeed, "collapsing" and "non-collapsing" states have been observed for Fe(100) [28].

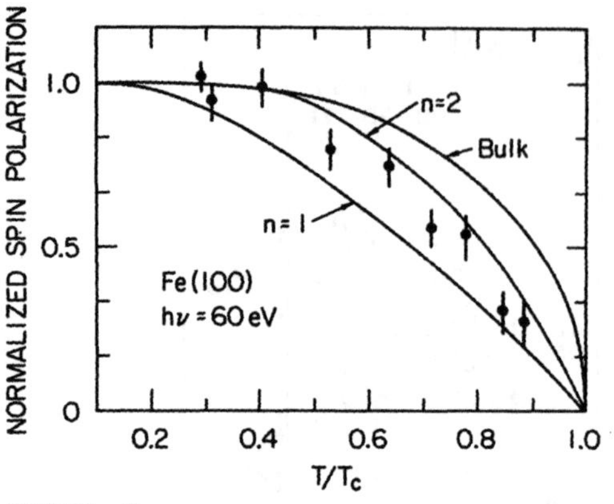

FIGURE 7.

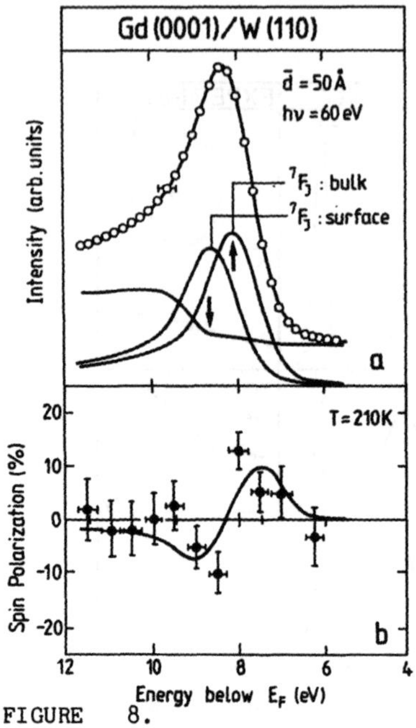

FIGURE 8.

The surface sensitivity for photoemission at photon energies near 60 eV is elucidated by measuring the spin polarization at constant binding energy as a function of temperature. If the electronic states probed do not shift in energy, it is obvious that the temperature dependence of the spin polarization P(T)/P(0) is equal to M(T)/M(0), where M(T) is the spontaneous magnetization [29]. For Fe(100), this applies at least approximately at E_b = 2.6 eV, the binding energy of $\Gamma'^{\uparrow}_{25}$ [16]. Measured values of P(T)/P(0) at 60 eV photon energy are shown in Figure 7. The data are compared with the calculation of the layer-dependence of M(T)/M(0) in the mean field approximation [29]. It is seen that the data points fall in between the curves calculated for the first and second layer. This corresponds to a probing depth of 1-3 layers.

3. Magnetic Surface Reconstruction.

Using synchrotron radiation, spin-polarized photoemission experiments on Gd(0001) reveal that the topmost atomic Gd-layer has a sizable antiferromagnetic coupling to the layers underneath. Figure 8 shows the intensity and spin polarization of the photoemitted electrons as well as the contributions from bulk and surface 4f-emission obtained by fitting the data with Doniach-Sunjic lines to account for the ^7F_j final-state multiplets (solid line). The different binding energy of the bulk and surface contributions is well known [30] and seems to be large enough (0.5 eV) to allow to detect the reversed polarization of the surface 4f-levels. In the frame of the simple semi-infinite model depicted in Figure 1 the experimental results indicate that the coupling constant J_{12} between next neighbors from the first and second atomic layers is negative. Figure 8 shows also the spin polarization calculated under the assumption that bulk and surface spins are anti-parallel (solid line).

4. Outlook.

We have demonstrated by case studies of a few examples the power of spin-polarized electron spectroscopy techniques for investigating magnetic phenomena. Future research will concentrate on topics which include the electronic structure of monolayer films and on more complicated systems such as Heusler and Invar alloys.

References

1. J. C. LeGuillou and J. Zinn-Justin, Phys. Rev. B 21, 3976 (1980).
2. R. Feder and H. Pleyer, Surf. Sci. 117, 285 (1982).
3. J. S. Reeve and A. J. Guttmann, Phys. Rev. Lett. 45, 1581 (1980).
4. A. J. Guttmann, G. M. Torrie, and S. G. Whittington, J. Magn. Magn. Mat. 15-18, 1091 (1980). β_1 is obtained from the values of β_1 given in References 3 and 4 by means of the scaling relations $\beta_1 + \gamma_1 = \beta + \gamma$.
5. H. W. Diehl and S. Dietrich, Z. Phys. B 42, 65 (1981).
6. K. Binder and D. P. Landau, Phys. Rev. Lett. 52, 318 (1984).
7. L. Onsager, Phys. Rev. 65, 117 (1944).
8. N. D. Mermin and H. Wagner, Phys. Rev. Lett. 17, 1133 (1966).
9. H. W. Diehl, "Field Theoretical Approach to Critical Phenomena at Surfaces" in Phase Transitions and Critical Phenomena, ed. by J. L. Lebowitz and C. Domb (Academic Press, London, to be published).
10. S. F. Alvarado, M. Campagna, and H. Hopster, Phys. Rev. Lett. 48, 51 (1982).
11. S. F. Alvarado, M. Campagna, F. Ciccacci, and H. Hopster, J. Appl. Phys. 53, 7920 (1982).
12. D. Weller, S. F. Alvarado, W. Gudat, K. Schröder, and M. Campagna, Phys. Rev. Lett. 54, 1555 (1985).
13. S. F. Alvarado, W. Eib, F. Meier, H. C. Siegmann, and P. Zürcher, in Photoemission and the Electronic Properties of Surfaces, ed. by B. Feuerbacher, B. Fitton, and P. F. Willis (John Wiley and Sons, Chichester, New York, Brisbane, Toronto, 1978).
14. R. Clauberg, W. Gudat, E. Kisker, E. Kuhlmann, and G. M. Rothberg, Phys. Rev. Lett. 47, 1314 (1981).
15. H. Hopster, R. Raue, G. Güntherodt, E. Kisker, R. Clauberg, and M. Campagna, Phys. Rev. Lett. 51, 829 (1983).
16. E. Kisker, K. Schröder, M. Campagna, and W. Gudat, Phys. Rev.Lett. 52, 2285 (1984); E. Kisker, K. Schröder, W. Gudat, and M. Campagna, Phys. Rev. B 31, 329 (1985).
17. V. Korenman, J. L. Murray, and R. E. Prange, Phys. Rev. B 16, 4032 (1977).
18. See, for example, B. L. Gyoffry, J. Kollar, A. J. Pindor, G. M. Stocks, J. Staunton, and H. Winter, in The Electronic Structure of Complex Systems, ed. by W. Temmerman and P. Phariseau, NATO Advanced Study Institute Series B, Vol. 113 (Plenum, New York, 1984).
19. T. Moriya, J. Magn. Magn. Mat. 31-34, 10 (1983).
20. V. Korenman and R. E. Prange, Phys. Rev. Lett. 53, 186 (1984).
21. E. M. Haines, V. Heine, and A. Ziegler, J. Phys. F 15, 661 (1985).
22. R. Feder, A. Rodriguez, U. Baier, and E. Kisker, Solid State Comm. 52, 57 (1984).
23. E. Kisker, J. Magn. Magn. Mat. 45, 23 (1984).
24. E. M. Haines, R. Clauberg, and R. Feder, Phys. Rev. Lett. 54, 932 (1985).
25. R. Clauberg, E. M. Haines, and R. Feder, Z. Phys. (in print).
26. H. Scheidt, M. Glöbl, V. Dose, and J. Kirschner, Phys. Rev. Lett. 51, 1688 (1983).
27. J. Unguris, A. Seiler, R. J. Celotta, D. T. Pierce, P. D. Johnson, and N. V. Smith, Phys. Rev. Lett. 49, 1047 (1982).
28. J. Kirschner, M. Glöbl, and H. Scheidt, Phys. Rev. Lett. 53, 612 (1984).
29. E. Kisker, J. Chem. Phys. 87, 3597 (1983).
30. R. Kammerer, J. Barth, F. Gerken, A. Flodstrom, and L. I. Johansson, Solid State Comm. 41, 435 (1982).

Magnetic Properties of Surfaces Investigated by Spin-Polarized Electron Beams

D.T. Pierce

National Bureau of Standards, Gaithersburg, MD 20899, USA

A spin-polarized electron beam incident on a ferromagnetic surface results in elastically and inelastically scattered electrons and in photons via radiative transitions. The spin-dependent intensities of each of these provide a sensitive measure of surface magnetization. A comparison between low-temperature spin deviations at the surface and in the bulk is given; the variation follows the same power law with temperature but with a larger prefactor for the surface. The connection between surface electronic structure and surface magnetism and the changes in each induced by chemisorption have been studied by spin-polarized inverse photoemission. For oxygen and carbon monoxide on Ni(110), a reduction of the Ni magnetic moment is found, rather than a decrease in exchange coupling and corresponding randomization of the alignment of the moments. Further, in the case of CO, the chemisorption interaction is nonlocal with one CO molecule eliminating on the average the magnetic moment of two Ni atoms.

1. Introduction.

The surface or solid-vacuum interface comprises a low-dimensional system with magnetic properties which may be different from the bulk of a material for a number of reasons. At the surface the bulk symmetry is broken. There is a lack of translational periodicity, a different number of neighboring atoms for the surface atoms, and possibly even a different arrangement of these atoms. Because the interface between a magnetic material and a second material has many similarities to the solid-vacuum interface, we expect that our studies of the latter will also be important in understanding atomically engineered epitaxial magnetic films, magnetic interfaces, and compositionally modulated (artificially layered) magnetic structures.

There are many questions about the magnetic properties at a surface which one would like to answer.

(1) Structure: What is the spatial variation of the magnetization from the surface into the bulk? Is there a layer-dependent magnetic moment? What is the magnetization variation parallel to the surface, i.e. the domain microstructure?

(2) Magnetization curves: Is the hysteresis loop at the surface different from the bulk? What does this tell us about the role of the surface in domain nucleation or magnetization rotation?

(3) Temperature effects: How do the low-temperature spin deviations at the surface differ from the bulk? How does the magnetization vary with temperature near the Curie temperature? What is the critical exponent? What is the extent of short range magnetic order above the Curie temperature?

(4) Electronic structure: How do the spin-split energy bands vary throughout the Brillouin zone? How are the core levels affected by exchange interactions with magnetic valence electrons?

(5) Chemistry: What magnetic and electronic properties change when a gas overlayer is adsorbed on the surface or when a species is segregated to the surface? What is the connection between changes in the magnetization and changes in the electronic structure?

(6) Excitations: What is the spin-wave dispersion at the surface? What is the spectrum of Stoner excitations?

Electron spectroscopies are widely used in surface studies because low-energy electrons are strongly interacting and have a short inelastic mean free path in the material. In contrast to bulk probes such as conventional neutron or x-ray measurements, optical techniques, or transmission electron microscopy, low-energy electrons probe only the outer few atomic layers. Since some surface disturbances "heal" within an electron probing depth, it is also possible to obtain information on bulk properties. Distinguishing surface and bulk effects is not always straightforward, and caution is necessary in interpreting the data. In certain instances, such as the addition of an overlayer, the effect of the perturbation is localized at the surface and the induced changes are readily observed.

Normally, electron spectroscopies measure electronic properties and are insensitive to magnetic properties. However, when the spin polarization of the electrons is added as a parameter in studies of systems with a net spin orientation, such as ferrimagnets or ferromagnets, information can be obtained about the size of the magnetic moment as well as the degree of magnetic order. The component along the z direction of the spin polarization vector $\vec{P}$ of an electron beam is defined as

$$P_Z = \frac{n_\uparrow - n_\downarrow}{n_\uparrow + n_\downarrow} , \qquad (1)$$

where $n_\uparrow$ and $n_\downarrow$ are respectively the number of electrons with electron-spin parallel or antiparallel to the z direction. In studies of magnetic surfaces, the preferred direction is given by the magnetization or by the orientation of the net spin density of the material.

The answers to our questions about the magnetic properties of surfaces are accessible using a variety of experimental techniques involving spin-polarized electrons. The techniques can be conveniently divided into three categories: (a) polarized electrons <u>in</u> and photon or electron intensities <u>out</u>, (b) photons or unpolarized electrons <u>in</u> and polarized electrons <u>out</u>, and (c) polarized electrons <u>in</u> and polarized electrons <u>out</u>.

In the first category are spin-polarized electron scattering and polarized low-energy electron diffraction (PLEED). These techniques have been used to investigate surface magnetization (hysteresis) curves [1] as well as low temperature spin deviations [2] (see Section 2) and critical behavior [3]. In principle it is possible to use PLEED to extract layer-dependent magnetization if measurements are made at sufficiently low temperature; analysis [4] of higher temperature data is difficult owing to thermal fluctuations of the spin orientations which tend to be comparable to or larger than the intrinsic variation to be determined. In addition to these elastic scattering examples, inelastic polarized electron scattering has been used to detect Stoner excitations [5]. Another example which can be considered a type of inelastic scattering is spin-polarized inverse photoemission (Section 3). In this technique polarized incident electrons lose energy through excitation of radiative transitions in the target, and the resulting photons are detected. Such measurements have been used to investigate spin dependent electronic structure [6,7]. They have been especially fruitful in studying the connection between chemisorption-induced changes in surface magnetism and electronic structure [8-10].

Experimental techniques in the second category (b) involve the measurement of the spin polarization of an electron beam. Spin-polarized photoemission is the oldest [11] such technique and has been used with great success to study spin-dependent electronic structure [12,13], effects of chemisorption [14], and even short-range magnetic order near the Curie temperature [15]. The first measurement of the energy spectrum [16] of the spin polarization of low-energy secondary electrons emitted from a ferromagnetic material showed an enhancement of the polarization at low energies which was subsequently observed on other materials [17-19] and discussed theoretically [20,21]. Some depth resolution is possible by varying the secondary emission parameters. This makes it possible to compare surface and bulk hysteresis curves [22]. When the secondary electrons are excited by a finely focused electron beam in a scanning electron microscope, the magnetic domain microstructure can be imaged [23,24]. Spin-polarized Auger spectroscopy, the newest of these techniques in category (b), provides information on spin-dependent electronic structure and surface chemistry, and has the advantage that element specific magnetic information can be obtained [25]. Further examples of fruitful applications of polarized electron spectroscopies can be found in a number of reviews [12,26,27].

The third category (c) of experiments, in which the incident electrons are spin polarized and spin analysis takes place after the interaction, combines the difficulty of the previous two categories. One experiment of this type has been reported in which Stoner excitations in iron are investigated by using low-energy inelastic scattering of a polarized electron beam with measurement of the polarization of the scattered electrons [28]. With sufficient energy resolution, such experiments will also be able to map surface magnon dispersion curves.

Many of the examples cited above take advantage of the possibility of also using electrons to investigate bulk properties. In this chapter we are concerned with magnetic properties of low-dimensional systems and will focus on some experimental results on the magnetic properties of surfaces using two techniques of category (a). In Section 2 we discuss spin-polarized elastic scattering studies of low-temperature surface spin deviations. In Section 3 we use spin-polarized inverse photoemission to determine the connection between surface magnetism and electronic structure and the changes induced by chemisorption of overlayers of oxygen and carbon monoxide on a Ni(110) surface.

2. Spin Polarized Electron Scattering.

A. Spin Dependent Interactions.

When a spin-polarized electron beam is scattered from the surface of a material, the scattered intensity may depend on the orientation of the spins of the incident electrons. The important terms of the interaction Hamiltonian can be written

$$H = \Sigma_i V(\vec{r}-\vec{r}_i) + \Sigma_i \frac{1}{2m^2c^2} \frac{1}{|\vec{r}-\vec{r}_i|} \frac{dV(\vec{r}-\vec{r}_i)}{dr} \vec{s} \cdot \vec{L} + \Sigma_i J(\vec{r}-\vec{r}_i) \vec{s} \cdot \vec{S}_i \quad , (2)$$

where $\vec{r}$ is the position of the incident electron with spin $\vec{s}$ and angular momentum $\vec{L}$, $\vec{r}_i$ is the position of the i-th atom with spin $\vec{S}_i$, and $J(\vec{r}-\vec{r}_i)$ is the exchange integral. The first term is the spin-independent scattering potential, the second is the spin-orbit scattering potential, and the third is the spin-dependent part of the exchange potential. The last term can be written in terms of the net spin density, i.e. $\vec{S}_i$ need not be a localized atomic spin; the spin dependent part of the exchange-correlation potential of VON BARTH and HEDIN [29] can be cast in this form [30].

By reversing the orientation of the incident electron spin $\vec{s}$ it is possible to isolate the spin-dependent scattering from the spin-independent part arising from the first term of (2). Further, if $\vec{s}$ lies in the scattering plane and therefore perpendicular to $\vec{L}$, one discriminates against the spin-orbit scattering. Alternatively, by reversing the magnetization and hence the direction $\vec{S}_i$ of the net spin density of the target, the spin dependence of the scattering that arises solely from the target surface magnetization can be isolated. As can be seen from (2), if there is a decrease in the magnetization either by the spins $\vec{S}_i$ becoming disordered or if the magnetic moment per atom (i.e. $\vec{S}_i$) decreases, it will be observed by the electron scattering.

We define a scattering asymmetry

$$A = \frac{I_{\uparrow} - I_{\downarrow}}{I_{\uparrow} + I_{\downarrow}} , \qquad (3)$$

where $I_{\uparrow}$ and $I_{\downarrow}$ correspond to the intensity of the scattered electrons when the incident electron beam is 100% spin polarized parallel and antiparallel, respectively, to the net spin density of the sample. For single scattering within the first Born approximation, it can be shown that the asymmetry is directly proportional to the magnetization [31,32],

$$A(\vec{K}) \propto \frac{J(\vec{K})}{V(\vec{K})} \frac{M(T)}{M(0)} , \qquad (4)$$

where $\vec{K}$ is the momentum transfer and $J(\vec{K})$ and $V(\vec{K})$ are the Fourier transforms of the exchange and Coulomb potentials. While A is frequently proportional to the magnetization, this is not true in general when multiple scattering is present [13]. Then we have

$$A = \alpha\, M(T) + \beta\, M(T)^3 + \gamma\, M(T)^5 + \ldots \qquad (5)$$

and special care must be taken in interpreting the scattering data.

B. The Spin Polarized Electron Source.

The key to spin-polarized electron scattering experiments, or the inverse-photoemission experiments of Section 3, is having an electron gun which produces spin-polarized electrons with a direction of polarization that can be easily reversed. Such a source of spin-polarized electrons is realized in an electron gun which uses negative electron affinity (NEA) GaAs as a photocathode [34]. The polarization of the electrons results from selection rules governing the photoexcitation of electrons with circularly polarized light from the spin-orbit-split valence band to the conduction band of GaAs. The theoretical spin polarization changes from +50% to -50% when the helicity of the light is reversed.

In addition to producing spin-polarized electrons, the GaAs photocathode, when treated with cesium and oxygen to obtain a negative electron affinity, is a very efficient photoemitter. Therefore, with lasers of only moderate power, it is possible to obtain very intense electron beams. In short, the NEA GaAs polarized-electron gun achieves the favorable characteristics of a conventional electron gun. Additionally it yields a spin-polarized electron beam in which the spin polarization is easily modulated. The time structure of the electron beam intensity is also easily controlled by controlling the time structure of the exciting laser radiation. The energy spread of the beam varies from about 200 meV, for a cathode activated for optimum intensity, down to a remarkable 30 meV for a cathode activated for minimum energy spread [35]. While crucial to the measurements we describe, the operating principles and characteristics of the spin-polarized electron gun have been described [34] in great detail elsewhere and will not be further discussed here.

C. Low Temperature Spin Deviations.

Deviations at finite temperature from the perfect alignment of the electron spins at 0 K are known to occur by thermal excitation of spin waves or magnons. The ratio of the bulk magnetization at a temperature T, $M_b(T)$, to the bulk magnetization at zero temperature $M_b(0)$, follows the Bloch law

$$M_b(T)/M_b(0) = 1 - B_b\, T^{3/2} + \ldots \qquad (6)$$

The constant B_b is characteristic of long-wavelength spin waves. The next higher

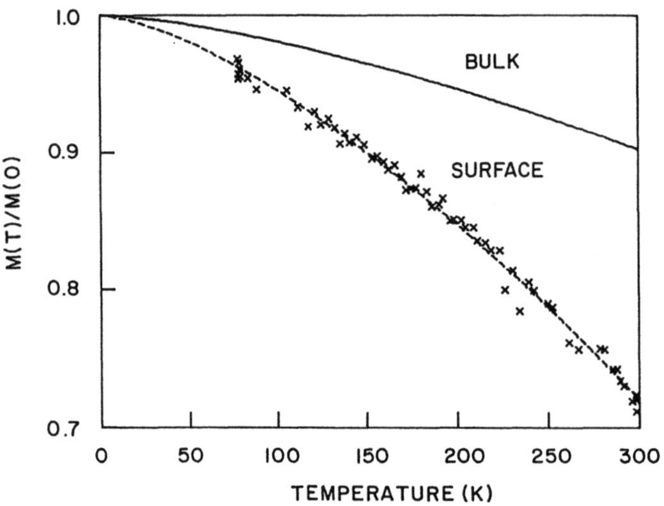

Figure 1. The temperature dependence of the surface magnetization (points) measured by polarized electron scattering is compared to that of the bulk (solid line) determined by conventional methods

order term, proportional to $T^{5/2}$, has been found to be small for amorphous ferromagnets at temperatures less than half the Curie temperature, T_c [36].

At the surface, the boundary conditions are different, and the deviation of the spontaneous magnetization from its zero-temperature value can be different from that in the bulk. Assuming that standing spin waves have antinodes at the surface, RADO [37] predicted in 1957 that the surface magnetization would decrease as $T^{3/2}$ but with a coefficient B_s twice the bulk value B_b. In a more extensive calculation, including both surface and bulk spin waves, MILLS and MARADUDIN [38,39] found that the surface spin-wave contribution is exactly cancelled by a reduction in the bulk spin-wave contribution, again leading to a $T^{3/2}$ temperature dependence with B_s twice the value of B_b. Although this result was obtained within the Heisenberg model, it is also expected to hold for itinerant systems at low temperatures where the spin waves have long wavelengths. These predictions of surface magnetic behavior remained untested for many years. The availability of a spin-polarized electron gun allowed the exploitation of the magnetic and surface sensitivity of polarized electron scattering to make such a test. Results are shown in Figure 1 for scattering from a ferromagnetic glass $Ni_{0.40}Fe_{0.40}B_{0.20}$ at an energy E = 90 eV and a scattering angle Θ = 166°. The applied field (remanent field of C-shaped electromagnet) was maintained at a constant value such that the sample was a single domain. The measured temperature dependence of the scattering asymmetry A(T) was shown to be proportional to the magnetization of the ferromagnetic glass [2]. The observed $T^{3/2}$ behavior represented by the dashed-line least-squares fit to the data from 80 K to 300 K was extrapolated to 0 K to compare with the relative bulk magnetization which was measured independently. If we plot the relative surface magnetization vs. the relative bulk magnetization at each temperature as in Figure 2, a straight line is obtained indicating that they both obey the same power law. The slope of the line, which is B_s/B_b, was found to be ~3. This value is a lower limit since the average electron probing depth was about 1.5 layers and some subsurface region was sampled.

The discrepancy between the experimental result and the theoretical prediction is attributed to the simplifications of the theory rather than a fundamental fault. The theory is for a simple cubic Heisenberg ferromagnet with uniform magnetic moments and exchange coupling. Extending the theory to allow, for example, for a reduced exchange coupling between the surface layer and the bulk compared to the bulk-bulk exchange coupling would lead to a larger B_s/B_b.

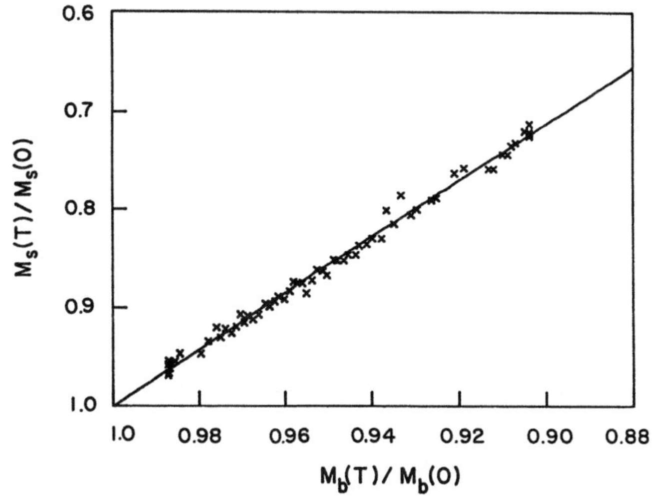

Figure 2. $M_s(T)/M_s(0)$ and $M_b(T)/M_b(0)$ have a linear relationship indicating that they vary with temperature according to the same power law. The slope of the line in $B_s/B_b \sim 3$.

3. Spin Polarized Inverse Photoemission.

A. The Technique.

In an ultraviolet photoemission experiment, a photon of known energy excites an electron from a filled state to an unfilled state above the vacuum level where it escapes into the vacuum. By measuring the kinetic energy of the electron, one then also knows the initial-state energy within the uncertainty of relaxation and screening effects. From the angle of emission, the parallel component of momentum k_{par}, which is conserved in the process, is also determined. Photoemission has in this way been able to map energy bands of materials and provide great insight into electronic properties [40].

In the approximately inverse process, a collimated electron beam is incident at a known energy and well defined angle (i.e. specific k_{par}) on the sample. If the electrons are of sufficiently low energy, any radiative transitions which occur will produce ultraviolet photons with negligible momentum. In this inverse photoemission process (also known as ultraviolet bremsstrahlung isochromat spectroscopy), the direct (vertical) transitions between bands allow one to map the energy bands of the unfilled states [41,42].

Shortly after the first k-resolved inverse photoemission measurements were demonstrated [43,44], it was possible, by replacing the conventional electron gun with a spin-polarized electron gun, to demonstrate [6] spin-polarized inverse photoemission spectroscopy (SPIPES). If the incident beam is spin polarized, the electron couples into an initial state of that particular spin orientation from which it can decay radiatively via a direct transition into an empty final state of the same spin. Thus it is possible to determine the spin-dependent band structure of a ferromagnetic material. In particular, in contrast to photoemission, with inverse photoemission it is possible to probe the energy range between the Fermi level and the vacuum level. Exchange-split unfilled d and f levels, which are important for ferromagnetism of materials, lie in this energy range.

The experimental apparatus is similar to that for electron scattering but instead of a Faraday cup to detect the electrons, a Geiger-Mueller counter detects the photons. The Geiger-Mueller counter [45] has a peak sensitivity at 9.7 eV and band pass of 0.7 eV determined by the combined effects of the iodine photoionization threshold and the ultraviolet transmission cutoff of the CaF_2 window. Advantages of such a detector are its simplicity and large solid angle of photons accepted. Spectra were obtained by varying the incident-electron energy and measuring at fixed photon energy. Although all SPIPES measurements to date have been made in this mode, spectrometers to select the photon energy will be used in the future.

B. A Ferromagnetic Surface: Ni(110).

The capability of SPIPES to investigate the electronic structure in the energy range between the Fermi energy and the vacuum level is used to advantage in studying the Ni(110) surface. A rather flat minority-spin d band of Σ_2 symmetry protrudes above the Fermi level as seen in the energy band diagram [6] of Figure 3. It is such minority-spin d holes that give rise to the average magnetic moment per atom in Ni of approximately a half Bohr magneton. Transitions to this d hole band can be detected very clearly as illustrated in the spectrum [46] of Figure 4. The spin-integrated (dashed curve) at an angle of incidence of 5° shows two peaks, but in the spin-resolved spectra at the bottom of the figure one observes three peaks, if the spin-split peak is counted as one. Transitions of the type A in Figure 3 give rise to the peak (1) in the minority spectrum. Exactly at normal incidence this transition is symmetry forbidden and grows in intensity at increasing angles of incidence.

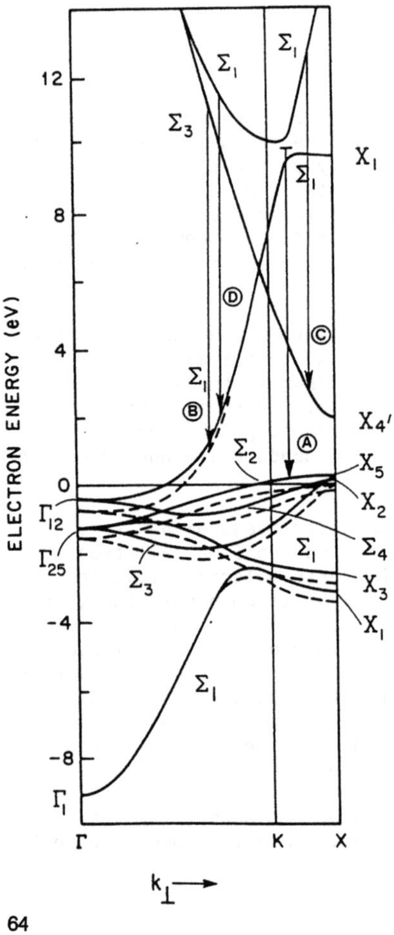

Figure 3. Band structure of Ni along the ΓKX direction showing energetically possible radiative transitions at $\hbar\omega = 9.7$ eV for electrons incident normally on the Ni(110) surface. Dashed curves distinguish the majority spin d bands.

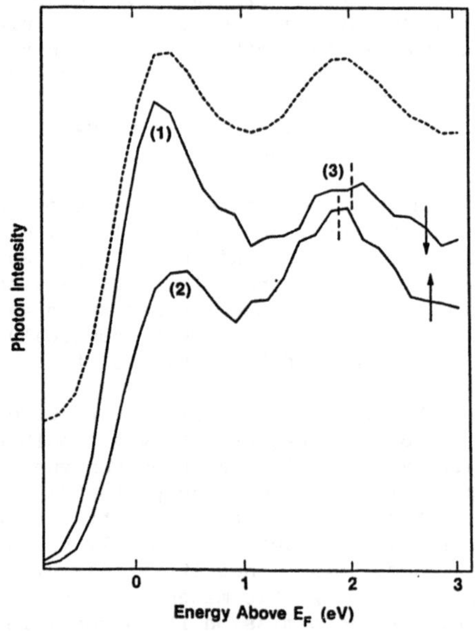

Figure 4. K-resolved spin-polarized (solid curves) and spin-integrated (dashed curve shifted upward) inverse photoemission spectra of Ni(110) are compared at an angle of incidence of 5°. The photon intensity is shown as a function of the final-state energy of the electron with respect to the Fermi level.

The origin of peak (2) in Figure 4 is different [46] from peak (1) as was seen from (i) its different energy position, (ii) its different dispersion and intensity variation with changing k_{par}, and (iii) its different behavior with oxygen chemisorption. It is not a majority spin replica of peak (1) such as would be observed if the sample were incompletely magnetized. Rather, peak (2) was attributed to the transitions of type (B) in Figure 3; the minority counterpart to peak (2) is obscured by peak (1) in the minority-spin spectrum.

The peak at ~2 eV above E_F is spin split. Further investigation of this peak showed that the splitting vanished when the sample was demagnetized, confirming its magnetic origin. This peak corresponds to a transition of type C in Figure 3. The splitting is not attributed to an exchange splitting of the s,p manifold itself but rather is an indirect consequence (via s-d hybridization) of the exchange splitting of the d band [46]. From second order perturbation theory, the s, p bands are shifted upward by an amount $H_{sd}^2/(E_s - E_d)$ where H_{sd} is the hybridization matrix element and E_s and E_d are the unperturbed band energies. Because of the exchange splitting of the d bands, the perturbation will have two values. This was the first observation of such a "differential" hybridization shift [46].

The energy bands of Figure 3 are calculated for electrons moving in bulk Ni normal to the (110) surface. With the estimated average 3-4 atomic layer probing depth of the SPIPES measurement for this surface, one would expect to see both bulk and surface features. In the absence of pronounced surface states or other perturbations at the surface, one cannot discern specifically surface features in the spectrum of Figure 4, which has a mixed surface and bulk origin. On the other hand, if there is a specific surface disturbance, such as a chemisorbed overlayer, the effect is clearly observed as described in the next section.

C. Chemisorption Induced Changes in Surface Magnetism.

The differences between the magnetic properties of a bulk ferromagnet and those of a low-dimensional system, such as a surface or interface, are intimately connected to differences in electronic structure. Traditionally, the relationship between surface magnetism and chemisorption has been explored using bulk magnetic measurements of high surface area samples [47]. In such measurements the connection to changes in electronic structure are not addressed. As discussed in the previous section, SPIPES allows one to distinguish transitions to particular final states, that is, to determine which electronic state or orbitals are actually involved in the chemisorption induced changes. We will focus on transitions to the minority-spin d hole states which give rise to the magnetic moment, and observe how these states change with chemisorption. Two quite different chemisorption systems, oxygen and carbon monoxide, will be discussed. The oxygen chemisorbs dissociatively, induces a 2x1 reconstruction of the Ni(110) surface, and sits in the long bridge site. Beyond initial rapid chemisorption at coverages of $\Theta \leq 0.5$ monolayer, there is nucleation and growth of NiO islands which are two to three layers thick [48]. Carbon monoxide, in contrast, chemisorbs molecularly, bonds predominantly in the terminal or on-top position, resides purely on the surface, and induces no reconstruction.

The role of the d electrons in chemisorption bonding is a topic of current discussion and dispute [49,50]. It has been suggested [49] that they are only weakly involved in the bonding, in which case the chemisorption may have only a weak and indirect effect on the surface electronic structure and surface magnetism as manifested by the d-hole spectrum. Alternatively, one might expect that there is a transfer of electrons to the electronegative oxygen which, depending on their relative spin orientation, might increase the surface magnetization. There are two distinct ways the surface magnetism can be decreased. First, without changing the size of the average moment, the chemisorption could cause a reduction of the exchange coupling in the surface resulting in the moments becoming randomly oriented. In this case one would still observe d-holes (i.e., the spin-integrated spectrum would be unchanged). However, there would be no preferred orientation, and a majority peak at the same energy as the minority-spin d-hole peak would be

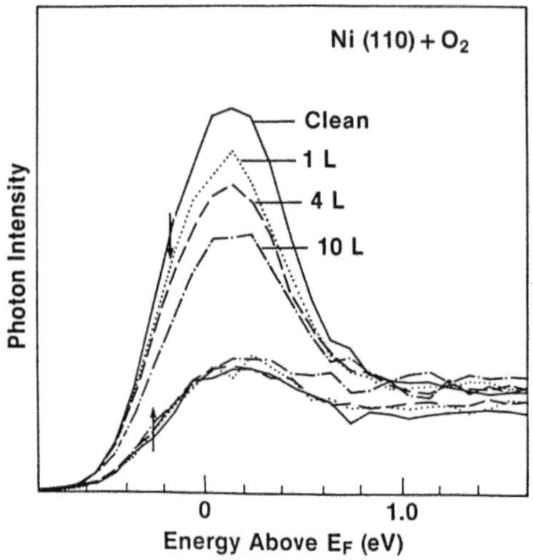

Figure 5. Spin-polarized inverse photoemission spectra, at an angle of incidence of 25°, of clean Ni(110) and after exposure to 1, 4, and 10 L of O_2. At each coverage two photon intensity spectra are shown, one for minority spin (↓) and the other for majority spin (↑) polarization of the incident electron beam.

produced by emission from surface atoms. Alternatively, if the chemisorption by some mechanism caused a reduction in the magnetic moment, i.e. in the number of d holes, one would only expect to see a decrease of the minority-spin peak of the SPIPES spectrum.

The effect of oxygen chemisorption is illustrated [8,9] by the spectra of Figure 5 for exposure of 1, 4, and 10 L (1 L = 10^{-6} Torr-sec) which corresponds to coverages of 0.3, 0.45, and 0.62 monolayer of oxygen. The major change that takes place at exposure up to 10 L is a reduction in the minority-spin d peak. There is no change in the small-majority spin peak which corresponds to transitions to a different final state; this indicates that electron attenuation in the overlayer cannot be an important cause for the decrease of the minority-spin d peak. Equally important as the reduction of the minority- spin peak is the absence of a growth of a corresponding majority-spin peak at the same energy as expected if the chemisorption just affects the exchange coupling and randomizes the moments. Nor is there any appearance of a majority- spin peak as might be expected if the chemisorption caused a decrease in the exchange splitting via a rigid shift of the bands.

These results from oxygen on Ni(110) can be compared to a similar investigation [51] of CO on Ni(110). As in the case of oxygen chemisorption, there is initially a smooth decrease of the minority-spin d peak with no concurrent growth of a majority-spin peak. [Unlike the case of oxygen, there was a slight but reproducible decrease in the majority-spin peak (2), but this change was minor and confined to the first 0.5 L of CO exposure.] Therefore, one knows that large decreases in the leading peak of the spin-integrated spectra result from the decrease in the minority-spin peak. Both the relative intensity of the leading peak in the spin-integrated spectra and the minority-spin spectra are plotted in Figure 6. Spin integrated spectra can be obtained faster with good statistics and for this reason were measured over a higher density of exposures and a wider range of final-state energies.

In contrast to the chemisorption of oxygen on Ni(110) where no oxygen antibonding peak was observed, a CO antibonding peak at 3.7 eV above the Fermi level was observed due to transitions into the CO Π* level. The relative intensity of this CO Π* peak is also shown as a function of coverage in Figure 6. The contribution to the spin-dependent photon asymmetry from transitions into the CO Π* level above the inelastic background was found [51] to be zero, A = 0 ± 0.02.

On comparing the behavior of the Ni minority-spin peak and the CO Π* peak in Figure 6, a major difference is noted. Whereas the CO peak increases uniformly

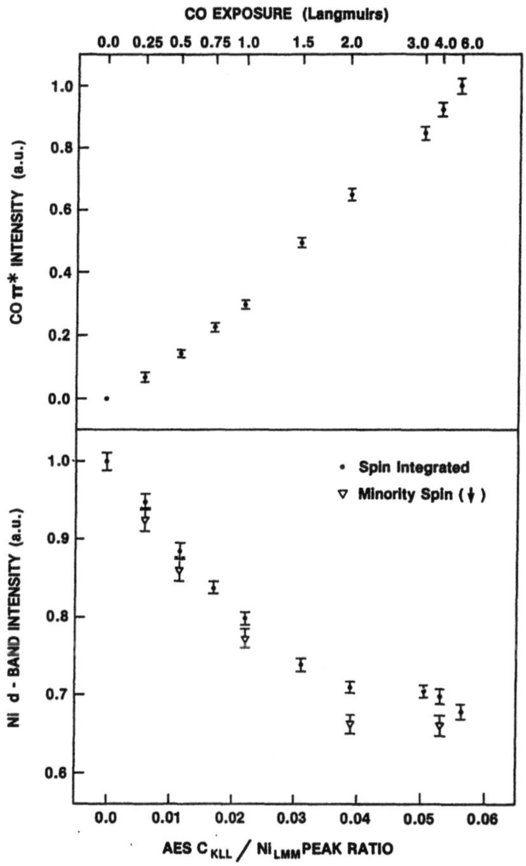

Figure 6. Intensity of spin-integrated inverse photoemission into the Ni d-band and CO Π* band, and spin-resolved inverse photoemission into the minority-spin band plotted as a function of coverage given by the AES C(272 eV)/Ni(848 eV) peak ratio. Error bars represent 2σ uncertainties from pulse counting statistics. Note that the CO intensity increases with coverage, whereas the decrease in the intensity of transitions to the d-band saturate at an AES ratio of ~0.03 which corresponds to 0.5 monolayer coverage.

with coverage, the Ni peak decreases and saturates at the Auger electron spectroscopy peak height ratio of 0.03 which is equivalent to 0.5 monolayer coverage [52]. There is no change in the surface magnetization after one CO molecule is adsorbed for each two Ni surface atoms. This result is in agreement with ferromagnetic resonance measurement of CO adsorption on thin Ni films where Göpel found that one CO effectively eliminated the surface magnetism of two Ni atoms. Such an effect is clearly nonlocal.

The nature of this nonlocal effect is clarified by the SPIPES measurement. As discussed in the case of oxygen, if the effect of chemisorption were to reduce the exchange coupling and randomize the orientation of the (still existing) moments, we would see the growth of a corresponding majority-spin d peak. Rather, the CO reduces the moment not only of the Ni atom involved in the chemisorption bond, but also of the Ni atoms in the vicinity of the CO adsorption site.

For both oxygen and carbon monoxide chemisorption on Ni(110) there is a substantial shifting of the density of states within the d-bands and even redistribution of state density between the d-bands. The SPIPES measurement has shown particular sensitivity to such changes. LIEBSCH [54] calculated for the related system, C(2x2)O/Ni(100), that the interaction of the O $2p_z$ states and the d-bands changed the density of states such that the minority-spin band of Ni(110) may be drawn below the Fermi level. A similar effect was found in calculations [55] of NiCO which show that the CO causes a change from a Ni d^9 to a d^{10} configuration. In another study of a larger cluster [56], Ni_{13} + CO, the unfilled minority-spin band was pushed below the Fermi level by CO chemisorption. The moments of all 13 Ni atoms in the cluster were changed clearly indicating a nonlocal effect. In other theoretical work [57,58], all-electron local-density functional slab calculations are now becoming available for C(2x2)O/Ni(100) and C(2x2)CO/Ni(100). SPIPES measurements of chemisorption on the Ni(100) surface will be able to test these calculations and some of the interesting predictions, such as a "spin-polarized" oxygen antibonding level on Ni(100) [57].

Acknowledgments.

The contributions to the work described in this chapter by my coworkers at NBS, R. J. Celotta, J. Unguris, C. S. Feigerle, and A. Seiler, and by visiting scientists H. C. Siegmann (ETH-Zurich) and J. L. Peña (Cinvestav-Mexico) are gratefully acknowledged. Thanks are also due to N. V. Smith who provided the Ni band structure and assisted in the analysis of the clean Ni(110) data. This work was supported in part by the Office of Naval Research.

References.

1. J. Unguris, D. T. Pierce, and R. J. Celotta, Phys. Rev. B 29, B81 (1984).
2. D. T. Pierce, R. J. Celotta, J. Unguris, and H. C. Siegmann, Phys. Rev. B 26, 2566 (1982).
3. S. F. Alvarado, M. Campagna, F. Ciccacci, and H. Hopster, J. Appl. Phys. 53, 7920 (1982).
4. R. Feder, S. F. Alvarado, E. Tamura, and E. Kisker, Surf. Sci. 127, 83 (1983).
5. J. Kirschner, D. Rebenstorff, and H. Ibach, Phys. Rev. Lett. 53, 698 (1984).
6. J. Unguris, A. Seiler, R. J. Celotta, D. T. Pierce, P. D. Johnson, and N. V. Smith, Phys. Rev. Lett. 49, 1047 (1982).
7. J. Kirschner, Surf. Sci. 162, 83 (1985).
8. A. Seiler, C. S. Feigerle, J. L. Peña, R. J. Celotta, and D. T. Pierce, J. Appl. Phys. 57, 3638 (1985).
9. C. S. Feigerle, A. Seiler, J. L. Peña, R. J. Celotta, and D. T. Pierce, J. Vac. Sci. Technol. A3, 1487 (1985).
10. D. T. Pierce, A. Seiler, C. S. Feigerle, J. L. Peña, and R. J. Celotta, J. Magn. and Magn. Mat., 54-57, 617 (1986).
11. G. Busch, M. Campagna, P. Cotti, and H. C. Siegmann, Phys. Rev. Lett. 22, 597 (1969).
12. H. C. Siegmann, F. Meier, M. Erbudak, and M. Landolt, Adv. Electron and Electron Phys. 62, 1 (1984).
13. E. Kisker, J. Phys. Chem. 87, 3597 (1983).
14. W. Schmitt, H. Hopster, and G. Güntherodt, Phys. Rev. B 31, 4035 (1985).
15. E. Kisker, K. Schröder, W. Gudat, and M. Campagna, Phys. Rev. B 31, 329 (1983).
16. J. Unguris, D. T. Pierce, A. Galejs, and R. J. Celotta, Phys. Rev. Lett. 49, 72 (1982).
17. E. Kisker, W. Gudat, and K. Schröder, Solid State Comm. 44, 623 (1982).
18. M. Landolt and D. Mauri, Phys. Rev. Lett. 49, 1783 (1982).
19. H. Hopster, R. Raue, E. Kisker, G. Güntherodt, and M. Campagna, Phys. Rev. Lett. 50, 71 (1983).
20. J. Glazer and E. Tosatti, Solid State Comm. 52, 905 (1984).
21. D. R. Penn, S. P. Apell, and S. M. Girvin, Phys. Rev. Lett. 55, 518 (1985).
22. R. Allenspach, M. Taborelli, M. Landolt, and H. C. Siegmann, Phys. Rev. Lett. 56, 953 (1986).
23. K. Koike and K. Hayakawa, Appl. Phys. Lett. 45, 585 (1985).
24. J. Unguris, G. G. Hembree, R. J. Celotta, and D. T. Pierce, J. Microscopy 139, RP1 (1985).
25. M. Landolt, R. Allenspach, and D. Mauri, J. Appl. Phys. 57, 3626 (1985).
26. J. Kirschner, Polarized Electrons at Surfaces (Springer-Verlag, Berlin-Heidelberg-New York-Tokyo, 1985).
27. R. Feder, Polarized Electrons in Surface Physics (World Scientific Publishing Co., Singapore, 1985).
28. J. Kirschner, Phys. Rev. Lett. 55, 972 (1985).
29. U. von Barth and L. Hedin, J. Phys. F 5, 1629 (1972).
30. D. T. Pierce and R. J. Celotta, Adv. Electron and Electron Phys. 56, 219 (1982).

31. X. I. Saldaña and J. S. Helman, Phys. Rev. B 16, 4978 (1977).
32. R. J. Celotta, D. T. Pierce, G.-C. Wang, S. D. Bader, and G. Felcher, Phys. Rev. Lett. 43, 728 (1979).
33. R. Feder and H. Pleyer, Surf. Sci. 117, 285 (1982).
34. D. T. Pierce, R. J. Celotta, G.-C. Wang, W. N. Unertl, A. Galejs, C. E. Kuyatt, and S. R. Mielczarek, Rev. Sci. Instrum. 51, 478 (1980).
35. C. S. Feigerle, D. T. Pierce, A. Seiler, and R. J. Celotta, Appl. Phys. Lett. 44, 866 (1984).
36. S. N. Kaul, Phys. Rev. B 24, 6550 (1981).
37. G. T. Rado, Bull. Am. Phys. Soc. II 2, 127 (1957).
38. D. L. Mills and A. A. Maradudin, J. Phys. Chem. Solids 28, 1855 (1967).
39. D. L. Mills, Comments Solid State Phys. 4, 28 (1971); 4, 95 (1972).
40. F. J. Himpsel, Adv. Phys. 32, 1 (1983).
41. V. Dose, J. Phys. Chem. 88, 1681 (1984).
42. N. V. Smith, Vacuum 33, 803 (1983).
43. D. R. Woodruff and N. V. Smith, Phys. Rev. Lett. 48, 283 (1982).
44. G. Denninger and V. Dose, Phys. Rev. Lett. 48, 279 (1982).
45. G. Denninger, V. Dose, and H. Scheidt, Appl. Phys. 18, 375 (1979).
46. A. Seiler, C. S. Feigerle, R. J. Celotta, D. T. Pierce, and N. V. Smith, private communication and to be published.
47. P. W. Selwood, Chemisorption and Magnetization (Academic, New York, 1975) and references therein.
48. P. H. Holloway, J. Vac. Sci. Technol. 18, 653 (1981).
49. C. F. Melius, Chem. Phys. Lett. 39, 287 (1976).
50. C. M. Varma and A. J. Wilson, Phys. Rev. B 22, 3795 (1980); A. J. Wilson and C. M. Varma, Phys. Rev. B 22, 3805 (1980).
51. C. S. Feigerle, A. Seiler, J. L. Peña, R. J. Celotta, and D. T. Pierce, Phys. Rev Lett. 56, 2207 (1986).
52. B. J. Bandy, M. A. Chesters, P. Hollins, J. Pritchard, and N. Sheppard, J. Mol. Spec. 80, 203 (1982).
53. W. Göpel, Surf. Sci. 85, 400 (1979).
54. A. Liebsch, Phys. Rev. B 17, 1653 (1978).
55. C. M. Kao and R. P. Messmer, Phys. Rev. B 31, 4835 (1985).
56. F. Raatz and D. R. Salahub, Surf. Sci. 146, L609 (1984).
57. S. Chubb and A. J. Freeman, to be published.
58. E. Wimmer, C. L. Fu, and A. J. Freeman, Phys. Rev. Lett. 55, 2618 (1985).

Surface Magneto-optic Kerr Effect Studies of p(1x1) Fe/Au(100)

*S.D. Bader, E.R. Moog, and P. Grünberg**

Materials Science and Technology Division, Argonne National Laboratory, Argonne, IL 60439 USA

Recent successful efforts to extend the conventional magneto-optic Kerr effect technique to the monolayer and submonolayer regime of Fe are summarized. Magnetic hysteresis data are presented for clean Fe films epitaxially deposited on Au(100) in the 1×1 surface structure. The Curie temperature for a 0.56-monolayer Fe film is substantial (150°C) but markedly reduced from that for bulk Fe. A layer of Au is found to cover the deposited Fe to lower the surface free energy.

1. Introduction.

Current activity in the area of epitaxial magnetic monolayers and thin films can be traced to recent developments in three related disciplines. Firstly, the experimental surface magnetism field, focused mainly on surfaces of bulk magnetic materials, has matured into a thriving area [1]. Secondly, the metal-on-metal epitaxy field has parallelled, to an extent, the progress of the corresponding semiconductor area [2]. Finally, computational physics efforts have helped guide experimental studies, due to advances in high-performance computer technology [3]. Questions of basic interest being addressed are concerned with the existence and type of surface magnetic order, and the size of the moment for itinerant magnetic systems. Controlling the coercivity and anisotropy are also very important for device applications. All of these issues intimately relate to dimensionality, geometry, and morphology. Thus, the growth conditions, epitaxy, and surface quality must be carefully characterized in order to interpret properly magnetic properties. In the present work we summarize our first efforts to utilize the surface magneto-optic Kerr effect (SMOKE) to study epitaxial films of p(1×1) Fe/Au(100) in the monolayer range [4,5].

2. Results.

An fcc Au(100) single crystal was chosen as the substrate in these studies because the primitive two-dimensional (2D) unit cell of its bulk-terminated surface is well matched to serve as a template for bcc Fe(100) growth. The lattice mismatch is found to be only 0.3% by comparing the (100) face in-plane spacing of 2.88 Å for Au and 2.87 Å for Fe. Although the clean Au(100) surface reconstructs to a 5×20 structure, deposition of only 15% of a monolayer of Fe is sufficient to stabilize the Au(100) 1×1 bulk-terminated surface structure. Low energy electron diffraction (LEED) studies performed in situ in a 10^{-11} Torr environment, utilizing an Fe evaporator configuration described elsewhere [4,5], were used to confirm the expected epitaxy. In addition, growth at low deposition rates (~0.15 Å/min.) with the substrate at a moderately elevated temperature (230°C) yielded the sharpest LEED spots, indicative of a highly ordered surface.

The Fe thickness, growth mode, and surface cleanliness were monitored by Auger electron spectroscopy, utilizing a double-pass cylindrical mirror analyzer. Due to the surface sensitivity of Auger spectroscopy, the gold substrate signal might be

---

* Permanent address: Institut für Festkörperforschung der Kernforschungsanlage, Jülich, D-5170 Jülich, West Germany.

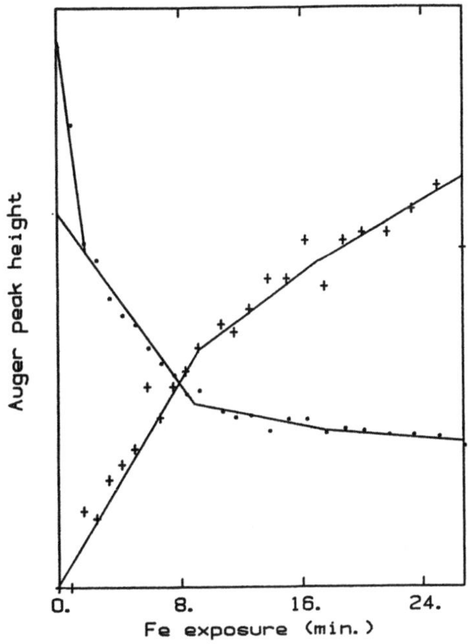

Fig. 1. Auger derivative signal peak-to-peak height as a function of Fe exposure. The dot symbol is for the Au (NVV) transition at 69 eV. The plus symbol is for the Fe (LMM) transition at 703 eV, enlarged four times. The initial rapid decrease in the Au signal spans the coverage range where the 5×20 Au reconstruction is reverting to a 1×1 LEED structure. The kinks at 9 and 18 min. indicate the growth is layer-by-layer. Monolayer Fe coverage is defined as being at the position of the 9-min. kink.

expected to exponentially vanish with increasing Fe dosage as the Fe signal asymptotically approaches a limiting value. For layer-by-layer growth linearly segmented exponentials are expected, due to each subsequent layer attenuating the signal from the layers below. Figure 1 shows such Auger data for Fe/Au(100). The clear break in slope at ~9 minutes Fe deposition is taken as the operational definition of monolayer coverage. Additional information is present in the figure at low and high Fe dosages. There is a slope change in the Au Auger signal at ~0.15 monolayer. This correlates with the structural transformation from the 5×20 to the 1×1 LEED pattern, and to a line-shape change [6] in the Au Auger signal associated with the two structures having different surface states, as is known from photoemission studies [7]. Also, for Cu(100), which does not reconstruct, this first kink is absent on Fe deposition [6]. At the highest Fe dosage the Au signal has not attenuated as much as would have been expected, based on mean-free-path estimates. At ~20 layers of Fe [Figure 2(a)] the Au Auger signal is still present with an intensity corresponding to monolayer coverage. This Au layer covers the Fe to lower the surface free energy. Proof of this, and that the Au does not intermix randomly throughout the Fe film, is that a light Ar^+-ion sputtering rapidly removes the Au. This is shown in Figure 2. Thus, we see that Auger electron spectroscopy provides much important information in addition to verifying the atomic purity of the surface region.

The SMOKE experimental arrangement is outlined in Figure 3. The Au(100) crystal is mounted between the poles of an electrosolenoid so that the magnetization direction is along the Fe[100] easy axis in the plane of the film. The He-Ne laser is polarized in the scattering plane, so that the longitudinal meridional Kerr effect is measured [8]. The linearly polarized incident light has its polarization rotated upon reflection from a ferromagnetic target due to the magneto-optic interaction. This can be envisaged as arising from the inequivalent interactions

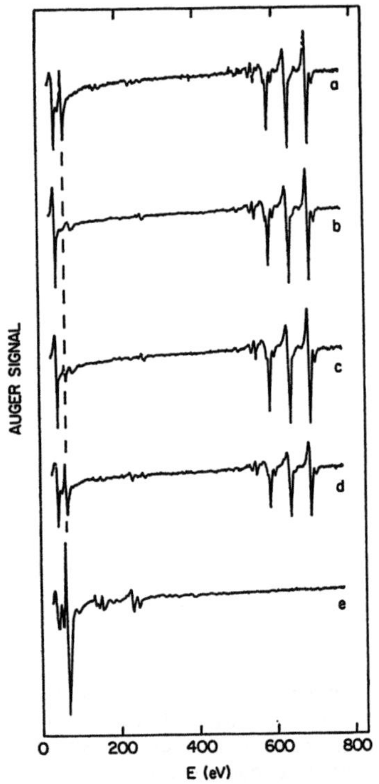

Fig. 2. Auger derivative spectra sputtering sequence. The dominant transitions are due to the Fe, except for the transition identified by the dashed vertical line due to Au. Spectrum (a) preceded the sputtering and is for ~20 layers of Fe deposited on Au(100), with an equilibrium adlayer of Au that rises to the surface. Spectra (b)-(e) were taken after 7, 55, 66, and 128 min. of sputtering, respectively. Note that the adlayer is rapidly removed as early as spectrum (b). A strong Au signal returns by spectrum (d) as the Fe film is sputtered away. Spectrum (e) is for the pure, annealled Au substrate after the sputtering is completed.

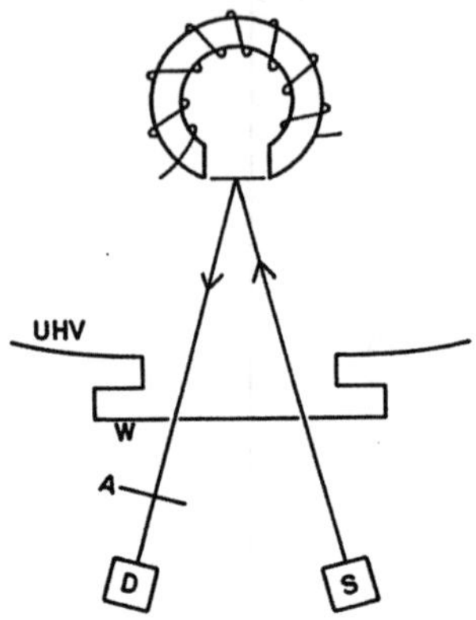

Fig. 3. The experimental agreement is shown schematically. An in-vacuum electromagnet is used to magnetize the sample. The polarized light from a laser source (S) shines through a window (W) onto the sample; the reflected light passes through the window and an analyzing polarizer (A) into the photodiode detector (D).

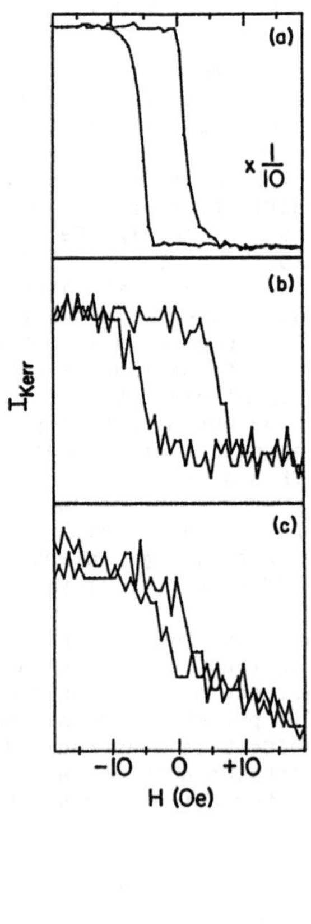

Fig. 4. Representative magnetic hysteresis curves are shown for (a) "thick" (>10 layers) Fe at 60°C (note height attenuation by a factor of 10 relative to the lower panels); (b) 0.56 layer of Fe at 65°C; and (c) 0.56 layers at 104°C.

experienced by the two oppositely-handed circularly- polarized components of the incident light [9]. The Kerr rotation of the light polarization vector causes the light intensity transmitted through the analyzer to change when the magnetization direction is reversed. This Kerr intensity I_{Kerr}, which is proportional to the sample magnetization, is superimposed on a large (10^2-10^3), constant background signal that is suppressed in the plots shown in Figure 4. Figure 4(a) shows a magnetic hysteresis curve for an Fe multilayer
(>10 layers thick) at 60°C, while Figs. 4(b) and (c) are for 0.56 monolayer Fe/Au(100) at 65° and 104°C, respectively. The hysteresis curves were obtained by signal averaging 40 to 80 scans over a 1 to 2 minute period. The squareness of the hysteresis loops is indicative of high-quality material. The reduction of the remanent magnetization and the coercivity H_c in Figure 4(c) relative to in Figure 4(b) indicates that bulk Brillouin-curve-type behavior is not obeyed and is suggestive of a reduced Curie temperature T_C for the submonolayer film. This is confirmed in Figure 5 where the remanent magnetization (plotted as a normalized I_{Kerr}) and the coercivity are plotted vs. temperature for four separate Fe dosages in the monolayer range. The T_C value for the 0.56-monolayer Fe coverage film is 150°C. For the higher Fe dosages the T_C values are above the upper limit of the experiments. This limit was set by the onset of bulk diffusion and/or desorption processes that cause the Fe Auger signal to decrease irreversibly with time above 300°C. For an Fe coverage of 0.4 monolayers (not shown in the figure) the T_C value is approximately room temperature. Note that the coercivity displays a stronger temperature dependence than I_{Kerr}, as expected, since H_c depends on a high power of the magnetization. The coercivity trends also suggest that T_C increases with Fe dosage.

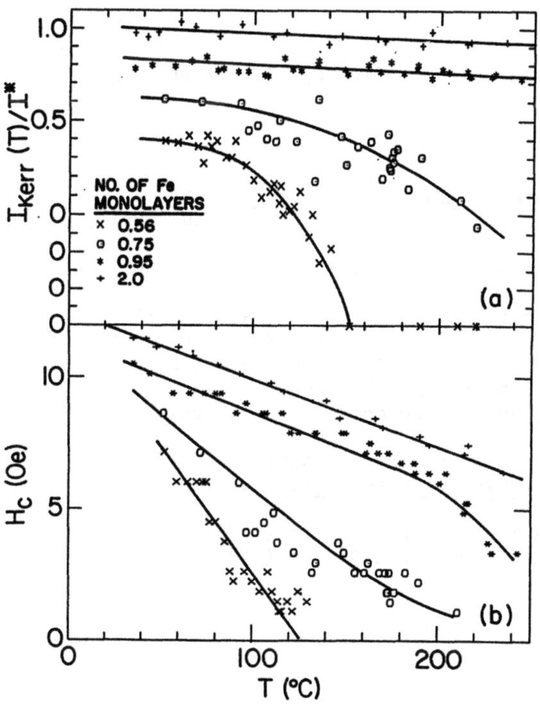

Fig. 5. Kerr intensity in remanence is plotted in (a), and the coercivity in (b) vs. temperature for different Fe depositions. The Kerr intensities for each deposition have been normalized by the near room temperature intensity I^* and then offset for clarity. The curves serve merely as guides to the eye. Data were taken on both heating and cooling cycles to ensure reproducibility of behavior.

3. <u>Discussion.</u>

Theoretical approaches, such as embodied in the mean-field theory and Monte Carlo simulations of BINDER and HOHENBERG [10], predict that T_C varies with the number of nearest neighbors, and decreases as the number of layers decrease. Thus, the T_C reduction reported above should, at first glance, be no surprise. However, two features of the present results are worth noting in relation to such theories. Firstly, the theories address free-standing, localized systems, while the present experiments were performed on an epitaxial, itinerant system. For the latter systems the magnetic moment in the monolayer range can be different than that for the bulk [3]. Secondly, the theories have not been extended to the submonolayer range, where the clearest evidence is presented herein for a T_C reduction.

To understand submonolayer magnetic phenomena the geometry must be specified. In the present case our LEED-Auger studies lead us to conclude that the Fe grows as a 2D layer covered by a layer of Au. The sharpness of the LEED spots suggests 2D island sizes at half-monolayer coverage in excess of ~50 Å. It is instructive to consider another common growth mode, in which isolated 3D islands form on top of the substrate. This type of growth has been reported to lead to superparamagnetic behavior for Fe [11]. Thus, the clear ferromagnetic signature observed in the present studies provides an auxiliary identification of the dominant growth mode that characterizes our films. Indeed, we have observed that altera- tions in the substrate temperature during deposition affect the type of film growth and the shape of the hysteresis curves observed in the SMOKE experiments [5].

There are two main findings we emphasize in closing. Firstly, the magneto-optic Kerr effect has been demonstrated for the first time to be applicable in the monolayer and submonolayer Fe regime. This gives rise to the acronym SMOKE and to a convenient-to-perform auxiliary probe to augment the increasing number of probes that are opening up rich opportunities in the field of surface magnetism. Secondly, the observation that T_C varies in the submonolayer regime should motivate theoretical simulations of the role of finite-size effects in such a geometry.

Acknowledgment.

This work was supported by the U.S. Department of Energy, BES-Materials Sciences, under Contract W-31-109-ENG-38.

References.

1. H. C. Siegmann, in Dynamical Phenomena at Surfaces, Interfaces and Superlattices, ed. by F. Nizzoli, K. H. Rieder, and R. F. Willis (Springer-Verlag, Berlin, 1985), p. 162.
2. G. Zajac, S. D. Bader, and R. J. Friddle, Phys. Rev. B 31, 4947 (1985), and references therein.
3. C. L. Fu, A. J. Freeman, and T. Oguchi, Phys. Rev. Lett. 54, 2700 (1985).
4. S. D. Bader, E. R. Moog, and P. Grüberg, J. Magn. Magn. Mat. (in press).
5. E. R. Moog and S. D. Bader, Superlattices and Microstructures (in press).
6. S. D. Bader et al., private communication (unpublished).
7. P. Heiman, J. Hermanson, H. Miosga, and H. Neddermeyer, Phys. Rev. Lett. 43, 1757 (1979).
8. M. Prutton, Thin Ferromagnetic Films (Butterworths, Washington, 1964), p. 134.
9. J. F. Dillon, Jr., in Magnetic Properties of Materials, ed. by J. Smit (McGraw-Hill, New York, 1971), p. 149.
10. K. Binder and P. C. Hohenberg, IEEE Trans. Magn. 12, 66 (1976).
11. G. Bayreuther and G. Lugert, J. Magn. Magn. Mat. 35, 50 (1983).

Photoelectron Spectroscopy Studies of Cr(001) Near-Surface Antiferromagnetism and Surface Ferromagnetism

L.E. Klebanoff, S.W. Robey, G. Liu+, and D.A. Shirley*

Materials and Molecular Research Division, Lawrence Berkeley Laboratory, and Departments of Chemistry and Physics, University of California, Berkeley, CA 94720, USA

Angle-resolved photoelectron spectroscopy (ARPES) studies of Cr(001) near-surface and surface electronic structure are reviewed. The near-surface energy band dispersions were investigated along the [010] direction parallel to the crystal surface. The periodicity of these band dispersions indicates that the valence electrons experience and self-consistently establish antiferromagnetism in the near-surface layers of Cr(001). Two surface-sensitive photoelectron peaks are observed in normal-emission ARPES. The spectral characteristics of these surface-related states are consistent with calculations that predict a ferromagnetic Cr(001) surface phase. The temperature dependence of one of these features is interpreted as evidence for a Cr(001) surface magnetic phase transition. An ARPES study of the Cr(001) 3s core level is also reviewed. The relative intensities of the exchange-split photoelectron components change as the ARPES spectra become more surface sensitive. Surface contamination suppresses this intensity variation and decreases the energy width of the 3s spectral profile. Almost all of the spectral observations are consistent with a surface-sensitive enhancement of the atomic 3d spin at the Cr(001) surface.

1. Introduction.

We review here recent angle-resolved photoelectron spectroscopy (ARPES) studies of Cr(001) surface [1-3] and near-surface [4] electronic structure. Prior to these experiments, the nature of the Cr(001) surface magnetization had attracted considerable theoretical interest [5-9]. Recently, more comprehensive calculations [10,11] have been made of Cr(001) surface electronic structure and magnetism. The comparison of the ARPES results with these theories provides insight into the Cr(001) surface and near-surface magnetism.

The nearly half-filled 3d band of Cr produces unique electronic and magnetic properties in the bulk metal. Neutron diffraction measurements [12] of Cr reveal the existence of a spin-density wave [13] (SDW) whose periodicity is incommensurate with the body-centered cubic (bcc) crystal lattice. The true incommensurate SDW ground state of bulk Cr is too complicated to be treated easily with band-structure theory. However, several calculations [14-17] have been made of a commensurate SDW state corresponding to perfect itinerant antiferromagnetism. An antiferromagnetic (AF) arrangement defines a simple cubic (sc) magnetic lattice in which the spin polarization at the corners and at the center of the bcc crystal lattice are of equal magnitude but opposite orientation. Consequently, the reciprocal space lattice is sc, and the associated AF Brillouin zone is a volume of simple cubic symmetry [17]. This differs fundamentally from the symmetry of the paramagnetic (PM) phase, for which the reciprocal-space lattice is face-centered cubic. The associated PM Brillouin zone is a regular rhombic dodecahedron with twice the volume of the AF Brillouin zone. The relationship between the AF and PM Brillouin zones is shown in Figure 1. A commensurate spin structure would render equivalent

*Present address: National Bureau of Standards, Gaithersburg, MD 20899.
+Permanent address: Department of Physics, Zhejiang University, Hangzhou, People's Republic of China.

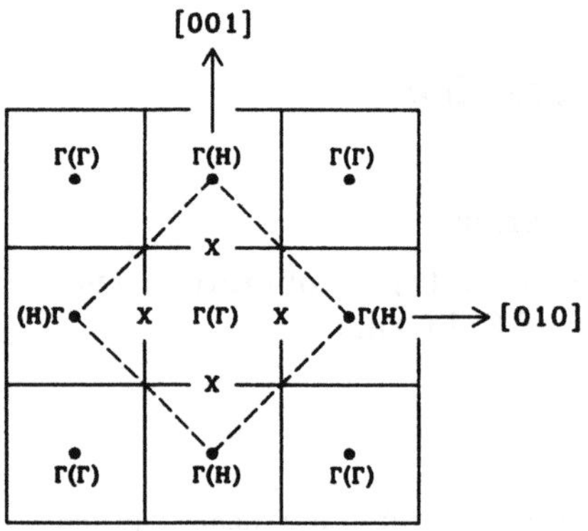

Figure 1. The AF (solid line) and PM (dashed line) Brillouin zones in the (100) mirror plane (defined by the [001] and [010] vectors). High symmetry points in the AF Brillouin zone are labelled with the conventional Greek letters; those in the PM Brillouin zone are labelled with the appropriate letters enclosed by parentheses.

the two symmetry points in reciprocal space labelled (Γ) and (H) in the PM Brillouin zone [17].

The existence of a magnetic lattice affects the periodicity of the valence-band dispersions. In the absence of antiferromagnetism, the bands would disperse with a periodicity consistent with the PM Brillouin zone. A magnetic potential with commensurate wave vector $2\pi/a$ would mix PM states with wave vectors k and $k \pm 2\pi/a$. Thus the energy-band structure for commensurate AF chromium arises [17] from the translation of the PM energy bands by $2\pi/a$. The periodicity of the valence-band dispersions would then be consistent with the sc Brillouin zone.

We expect from these considerations that the energy bands of the ideal AF state would disperse symmetrically about the X point of the sc Brillouin zone in Figure 1. In a PM state, the Cr valence bands would disperse about the (H) point, since this point corresponds to a zone boundary in the PM Brillouin zone. The marked difference between the AF and PM Brillouin zones in Figure 1 suggests that antiferromagnetism might be distinguishable from paramagnetism in chromium by a measurement of the valence-band dispersions along the [001] or [010] axes of Figure 1.

ARPES has proved to be a powerful probe of valence-band dispersion relations. Most ARPES experiments probe within the first ~10 atomic layers near a metal surface. Despite this surface sensitivity, for many metals semiquantitative agreement exists between ARPES results and bulk valence-band theory. This suggests that in these metals, the electronic structure is already "bulklike" by the second or third layer. However, Allan predicted [6-8] that the reduced coordination number (4) for the Cr(001) surface atoms would produce energy-band narrowing resulting in the formation of an unusual Cr(001) surface magnetic order. His tight-binding calculation [8] for Cr(001) predicted a ferromagnetic surface phase characterized by an exchange-split surface spin density of states (SSDOS), and large (2.8 μ_B) localized surface magnetic moments. The magnitude of the Cr(001) surface moment is thought to be much larger than the maximum value 0.59 μ_B found [12] for the bulk chromium magnetic moment.

Teraoka and Kanamori predicted [5] that Cr(001) surface magnetism could persist above the bulk Cr Neel temperature T_N of 312 K. Grempel extended the finite temperature theory using a spin-fluctuation formalism [9]. His results predict the persistence of Cr(001) surface ferromagnetic order for temperatures up to 850 K. The transition from surface to bulk properties likely encompasses a finite number of near-surface layers. It is the surface and near-surface regions that are probed by the ARPES measurements summarized here. We review in Section 2 our ARPES investigation of the Cr(001) near-surface valence electronic structure and its implications for near-surface magnetism. In Section 3 we review our photoemission study of the Cr(001) surface valence electronic structure. Section 4 presents core-level photoelectron spectroscopy data for the atomic Cr(001) surface electronic structure. Finally a summary of the ARPES results is presented in Section 5.

2. Cr(001) Near-Surface Electronic Structure and Antiferromagnetism.

Figure 2 displays five ARPES spectra out of 50 that were taken of Cr(001) (at 298 K) to probe the [010] direction for binding energy (E_{IN}) values of 2-4 eV. The value $k_{par}{}^{f}$ ($\AA^{-1}$) is the magnitude of the parallel component (along [010]) of the final-state wave vector for the peak marked with a tick. We use the direct-transition model [4] to associate $k_{par}{}^{f}$ with an initial-state wave vector $\vec{k}_i$ along the [010] symmetry line in the first Brillouin zone. For the PM Brillouin zone of Figure 1, $k_{par}{}^{f} = 0$ probes $\vec{k}_i$ at the H point. When viewed in the AF Brillouin zone of Figure 1, direct transitions to $k_{par}{}^{f} = 0$ arise from $\vec{k}_i$ at the Γ point.

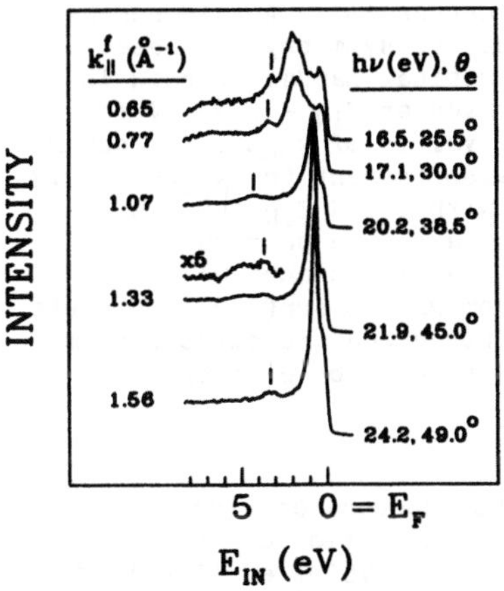

Figure 2. ARPES spectra of Cr(001) at 298 K. The photon energy hν (eV) and the polar angle of electron detection θ_e (degrees), measured from the surface normal, are listed for each spectrum. The values $k_{par}{}^{f}$ ($\AA^{-1}$) are the components of the final-state wave vectors parallel to the surface for the peaks labelled with a tick mark. The intensities of the spectra have been scaled to clarify the presentation.

Figure 3 plots E_{IN} versus $k_{par}{}^{f}$ for the features observed in the ARPES spectra. Three distinct bands of points are present. Band III is a feature that has been assigned [2] to the Cr(001) surface electronic structure. Band II is located at $E_{IN} = 2.4(1)$ eV for $k_{par}{}^{f} = 0.15(8)$ $\AA^{-1}$ in Figure 3. As $k_{par}{}^{f}$ increases, Band II disperses toward E_F. At $k_{par}{}^{f} \sim 1.06(8)$ $\AA^{-1}$, Band II suddenly flattens out, with $E_{IN} = 1.5(1)$ eV. Band I is located at $E_{IN} = 3.2(1)$ eV for $k_{par}{}^{f} = 0.20(8)$ $\AA^{-1}$. As $k_{par}{}^{f}$ increases, E_{IN} increases to a maximum of 4.1(1) eV at $k_{par}{}^{f}$ 1.07(8) $\AA^{-1}$. This value of $k_{par}{}^{f}$ corresponds to the X point (1.09 $\AA^{-1}$) of the theoretical AF Brillouin zone (Figure 1) to within the experimental uncertainty. As $k_{par}{}^{f}$ increases beyond 1.07(8) $\AA^{-1}$, Band I disperses symmetrically back toward E_F. In contrast to Band III, Bands I and II show comparatively little sensitivity to surface contamination. This suggests that Bands I and II arise from the Cr(001) near-surface electronic structure.

The smooth symmetrical dispersion of Band I about the X point in Figure 3 is strong evidence that the X point is a Brillouin zone boundary for the [010] near-surface electronic structure. This periodicity is consistent with that expected for an AF electronic structure. The periodicity of Band I's dispersion is certainly inconsistent with our expectations for a PM phase. We therefore interpret the dispersion of Band I in Figure 3 as evidence that the near-surface valence electrons feel, and self-consistently establish, antiferromagnetism in the near-surface layers of Cr(001). This conclusion is also consistent with the flattening of Band II at the X point, although the absence of symmetrical dispersion for Band II is not understood. These properties confirm the early theoretical predictions [5-9] that the Cr(001) near-surface region is antiferromagnetic. The temperature dependence of this near-surface antiferromagnetic order is reported in Reference 4. Those results indicate that Cr(001) near-surface antiferromagnetism persists up to 2.5 times T_N.

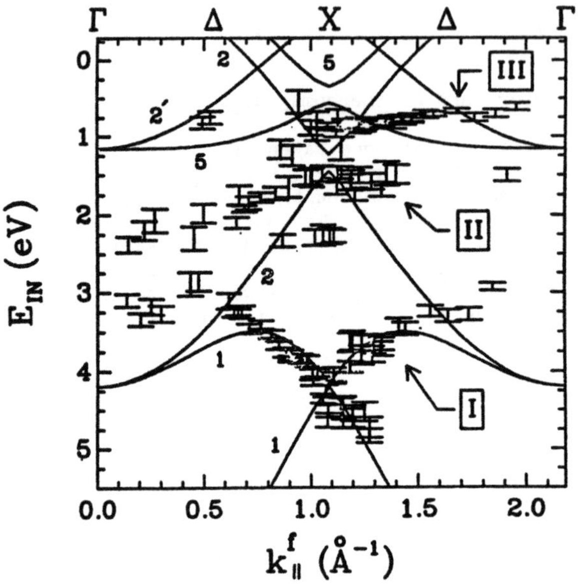

Figure 3. Experimentally observed band dispersions along the [010] symmetry line. Values of E_{IN} (eV) are plotted versus k_{par}^{f} ($Å^{-1}$) for the features observed in our ARPES spectra. The size of the symbol gives the experimental error in E_{IN}. The Greek letters at the top of the figure label the [010] symmetry line in the manner appropriate for simple cubic symmetry. The solid lines are the Γ-X-Γ band dispersions for commensurate AF chromium as calculated by Skriver (Reference 15). Arabic numerals label the symmetry of these theoretical bands. A peak is observed at E_{IN} ~ 0.25 eV for many values of k_{par}^{f}. A feature is also present for $k_{par}^{f} < -0.3$ $Å^{-1}$ with $0.7 \leq E_{IN} \leq 0.9$ eV. Both have been previously assigned (Reference 2) to the ferromagnetic surface electronic structure and are omitted from this figure for clarity.

3. Cr(001) Surface Electronic Structure and Magnetism.

Figure 4 compares room-temperature normal-emission ARPES spectra of clean Cr(001) and Cr(001) exposed to 5 L of CO [19]. Note the sharp attenuation of those features with binding energies 0.16(5) eV (labelled 1) and 0.75(5) eV (labelled 2). The surface sensitivity of the features 1 and 2 in Figure 4 indicates that these photoelectron peaks arise from initial states in the Cr(001) surface electronic structure. Since the polar angle of electron detection θ_e is zero, $k_{par}^{f} = 0$ $Å^{-1}$ for all photoelectron peaks in the spectra of Figure 4. Features 1 and 2 therefore arise from initial states at the $\bar{\Gamma}$ point of the Cr(001) surface Brillouin zone. Polarization-dependent ARPES measurements [2] have shown that the surface features 1 and 2 possess Δ_1 and Δ_5 symmetries, respectively.

Surface feature 2, and to a lesser extent feature 1, were observed to display a temperature dependence. The binding energies of 1 and 2, $E_{IN}(1)$ and $E_{IN}(2)$, are plotted against temperature in Figure 5. The thermal modification of $E_{IN}(2)$ is larger in magnitude and different in character than that of $E_{IN}(1)$. We have scaled and graphed in Figure 5 the temperature-dependent bulk exchange splitting in nickel Δ_{ex}^{Ni} as measured by ARPES [20]. Note that $E_{IN}(2)$ decreases with temperature in a manner very similar to that of a ferromagnetic exchange splitting.

The comparison of these ARPES results with theory provides insight into the magnetic properties of the Cr(001) surface. The observed characteristics of features 1 and 2 have found overall agreement with a comprehensive calculation of Cr(001) surface electronic structure recently completed by VICTORA and FALICOV [10]. In agreement with ALLAN [6-8] and GREMPEL [9], Victora and Falicov predict a ferromagnetic Cr(001) surface characterized by a very large (~3.00 electrons) surface spin polarization. Their theoretical prediction for the Cr(001) surface and near-surface magnetization is portrayed in Figure 6. Victora and Falicov also predict the symmetry-, wave vector-, and layer-dependence of the initial states contributing to the surface electronic structure. When discussing the layer dependence of the Cr(001) surface electronic structure, the terms "majority spin" and "minority spin" become ambiguous. The label (+)-spin is used in reference to electrons whose spin magnetic moments lie parallel to the magnetization of the

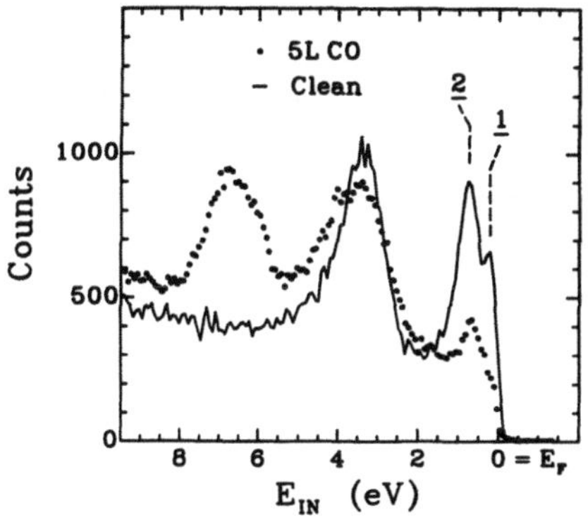

Figure 4. Normal-emission P-polarization ARPES spectra ($h\nu$ = 23.00 eV) of Cr(001) before (line) and after (dots) 5L CO exposure. The contamination-induced peak at 6.7 eV binding energy is assigned to impurity (carbon and oxygen) 2p photoemission.

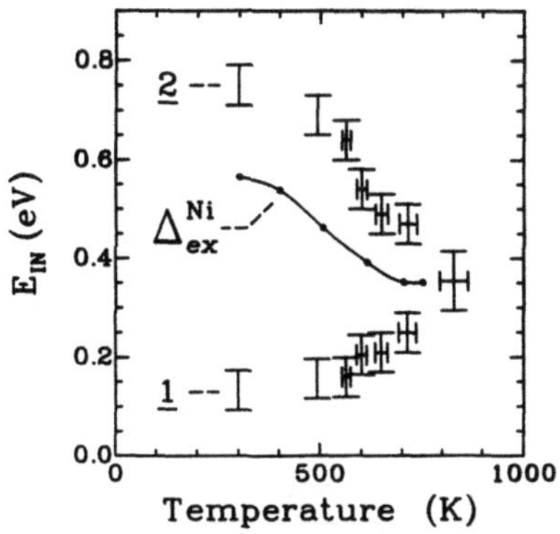

Figure 5. The binding energies of the surface features 1 and 2 plotted vs. temperature. The solid line is the temperature dependence of the bulk exchange splitting in nickel as measured by ARPES. This curve, reproduced from Figure 2 of Reference 20 and scaled in energy for comparison, should only be compared with the temperature dependence of feature 2.

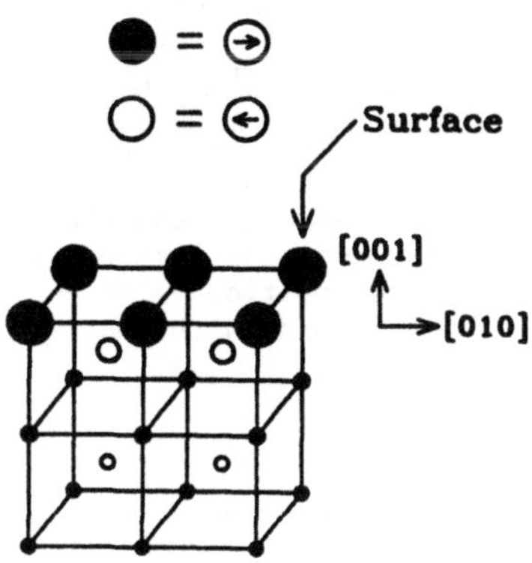

Figure 6. The Cr(001) surface and near-surface magnetization as calculated by Victora and Falicov [10]. Atoms whose spins point to the right are indicated by darkened circles. Atoms whose spins point to the left are symbolized by open circles. The diameter of the circle representing an atom is drawn proportional to the magnitude of the atom's spin polarization. The surface, second, and third layer spin, polarizations are predicted to be 3.00 electrons, -1.56, and 1.00 electrons, respectively.

surface layer. Electrons whose moments lie antiparallel to the surface magnetization (but parallel to the second-layer magnetization) are identified as (-)-spin electrons.

The theoretical Cr(001) surface-layer and second-layer electronic structures of $k_{par}^{f} = 0$ ($\bar{\Gamma}$) are shown in Figures 7(a) and 7(b), respectively. A prominent Δ_5-symmetry (+)-spin surface state is predicted with energy 1.30 eV below (to the left of) E_F in Figure 7(a). This is accompanied by a Δ_5-symmetry (-)-spin surface state with binding energy 1.2 eV in Figure 7(b). Both surface states are located in a Δ_5-symmetry band gap of the surface-projected theoretical AF band structure [14-17]. A small Δ_1-symmetry (+)-spin surface state is also predicted with energy 0.68 eV above (to the right of) E_F in Figure 7(a).

These aspects of the theoretical Cr(001) surface electronic structure at $\bar{\Gamma}$ are consistent with the ARPES results. The nearly degenerate Δ_5-symmetry theoretical surface states in Figures 7(a) and 7(b) have binding energies and dispersions similar to that observed for the Δ_5-symmetry surface feature 2. We therefore assign surface feature 2 to nearly degenerate Δ_5-symmetry (+)-spin and (-)-spin surface states at $\bar{\Gamma}$. As such, feature 2 would have d_{xz} and d_{yz} orbital character.

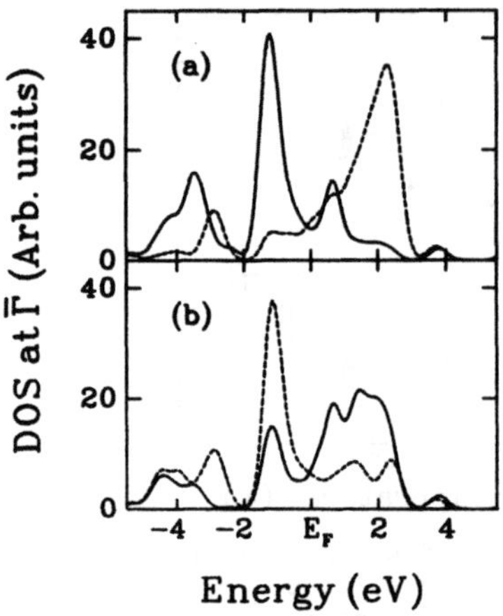

Figure 7. The theoretical Cr(001) surface-layer (a) and second-layer (b) electronic structure at $\bar{\Gamma}$, as calculated by Victora and Falicov [2]. For both (a) and (b) the (+)-spin surface electronic structure is indicated by the solid line; the (-)-spin surface electronic structure is drawn with the dashed line. The theory has been broadened in energy by a 0.6 eV Gaussian to simulate experiment. States to the left of E_F are occupied.

The surface feature 1 possesses Δ_1-symmetry. The only theoretical Δ_1-symmetry initial state near E_F is the unoccupied (+)-spin surface state with energy 0.68 eV above E_F in Figure 7(a). Note that a companion Δ_1-symmetry (-)-spin surface state is not found in the second-layer electronic structure, Figure 7(b). The true energy position of the (+)-spin surface state may be closer to E_F than predicted. This possibility, combined with the very sharp onset of the feature 1 spectral intensity at E_F, leads us to assign the Δ_1-symmetry surface feature 1 to a peaked spectral profile produced by the truncation of a Δ_1-symmetry (+)-spin surface state by the Fermi level. As such, feature 1 would have d_{z^2} orbital character. The overall agreement between theory [6-11] and the ARPES data indicate that the Cr(001) surface is in fact ferromagnetic. This interpretation, and the resemblance of the temperature dependence of feature 2 to $\Delta_{ex}^{Ni}(T)$ suggest that the temperature dependence of $E_{IN}(2)$ signals a Cr(001) surface magnetic phase transition. The observed temperature dependence of $E_{IN}(2)$ is in qualitative agreement with that recently predicted by HASEGAWA [18].

4. Cr(001) Atomic Surface Electronic Structure and Magnetism.

Core-level ARPES can be used to probe the valence-electron spin at an atom in a metal. Consider photoelectron emission from the atomic 3s level of a magnetic solid [21-26]. The simplest expectations [21-27] are of two photoelectron peaks. One peak corresponds to the high-spin final state <HS| in which the spin angular momentum of the remaining 3s electron is oriented parallel to the net final-state 3d spin vector $\vec{S}^f_{3d}$ at the photoionized atom. The second peak corresponds to the low-spin final state <LS| for which the spin of the remaining 3s electron is oriented antiparallel to $\vec{S}^f_{3d}$. The <HS| energy E(<HS|) is preferentially lowered below E(<LS|) by the intra-atomic 3s-3d exchange interaction.

One-electron theory [22-24,27] predicts that the "multiplet splitting" ΔE between E(<HS|) and E(<LS|) is given by:

$$\Delta E = E(\langle LS|) - E(\langle HS|) = (2S+1)\, K(3s,3d). \qquad (1)$$

In (1), K(3s,3d) is the intra-atomic exchange integral [23] between the 3s and 3d orbitals, and S is the initial-state atomic spin quantum number. The ratio R of the <LS| spectral intensity to the <HS| intensity is predicted [22,23,28] to be the ratio of the spin multiplicities for <LS| and <HS|:

$$R = \frac{2(S-1/2)+1}{2(S+1/2)+1} = \frac{S}{S+1}\ . \qquad (2)$$

Photoemission studies [21-26] of the 3s level of 3d magnetic metals and their compounds have qualitatively confirmed the above expectations. Intra-shell electron correlation can reduce the observed values of ΔE and R from the values predicted by these equations [29-31]. However, for a given element, ΔE and R vary with S in a manner qualitatively consistent [21-26] with (1) and (2). These facts suggest that an increase in the atomic 3d spin at the Cr(001) surface may manifest itself as a surface dependence of the Cr(001) 3s photoemission lineshape.

Figure 8 displays a normal-emission (θ_e = 0) 3s ARPES spectrum of clean Cr(001) using a photon energy of 156.0 eV. As observed previously [24], the Cr 3s photoelectron lineshape consists of a large component (referred to hereafter as component A) with binding energy 74.0(2) eV, and a smaller shoulder (referred to hereafter as component B) on the high binding energy side of component A. Since the kinetic energy of the 3s photoelectrons is ~75 eV, only the first ~5 Å or ~5 atomic layers [32] contribute substantially to the 3s photoemission spectrum of Figure 8.

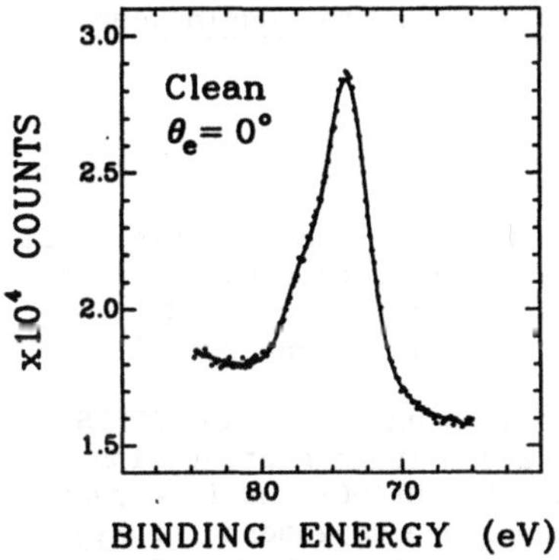

Figure 8. Normal-emission ($\theta_e = 0^\circ$) 3s ARPES spectrum (dots) of clean Cr(001). The solid line is a simplex fit to the spectrum (see Reference 3).

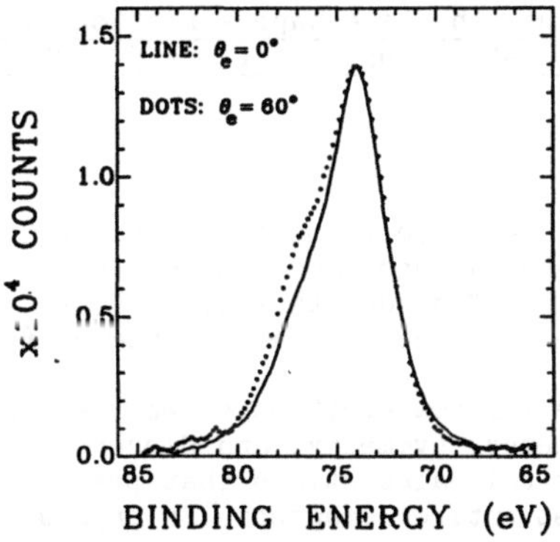

Figure 9. A direct comparison of $\theta_e = 0^\circ$ (line) and $\theta_e = 60^\circ$ (dots) 3s ARPES spectra of clean Cr(001). For each spectrum we have subtracted a quadratic background determined from simplex fits to the spectra (see Reference 3). The two background-subtracted spectra were smoothed with a 9-point least-squares quadratic convolution, and normalized to each other at 74.0 eV binding energy.

The surface sensitivity of our ARPES measurement can be enhanced by increasing the polar angle of electron detection θ_e. In Figure 9, we contrast the lineshape of a $\theta_e = 0^\circ$ spectrum (line) and a more surface sensitive $\theta_e = 60^\circ$ spectrum (dots). It is clear from Figure 9 that increasing the surface sensitivity of the ARPES measurement from ~5 Å ($\theta_e = 0^\circ$) to ~3 Å ($\theta_e = 60^\circ$) increases the ratio R of the component B intensity to that of component A. From (2) the observed increase in R indicates that the average atomic spin S detected within the first ~3 layers is larger than that detected within the first ~5 layers. This conclusion is qualitatively consistent with the magnetization profile of Figure 6. However, by (1), one would expect an increase in R to be accompanied by an increase in the overall width of the Cr(001) 3s lineshape. Such a broadening is not apparent in Figure 9.

Figure 10 displays the chemical sensitivity of the $\theta_e = 60^\circ$ Cr(001) 3s photoemission lineshape. After the surface is contaminated with ~0.5 monolayers of dissociated CO, both the observed value of R and the energy width of the 3s envelope decrease. This sensitivity is entirely consistent with the existence of

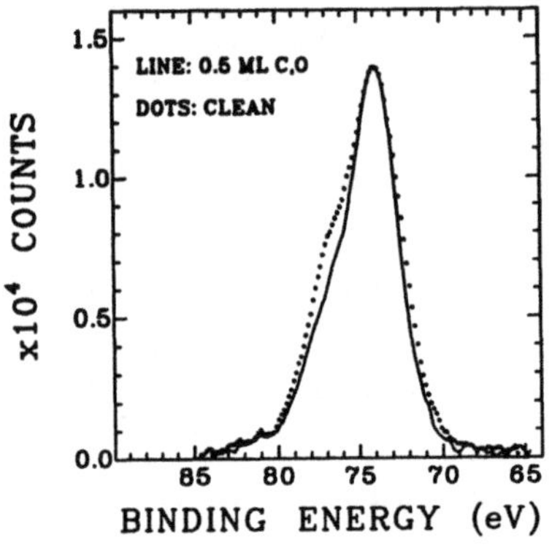

Figure 10. A direct comparison of the clean (dots) and contaminated (line) Cr(001) 3s ARPES spectra measured at $\theta_e = 60^\circ$. For each spectrum, we have subtracted the quadratic background obtained from simplex fits to the spectra (see Reference 3). The two background-subtracted spectra were smoothed and normalized as in Figure 9.

an enhanced atomic spin polarization at the clean Cr(001) surface. From (1) and (2), we attribute the lineshape variation of Fig. 10 to a contamination-induced decrease of the atomic spin S at the Cr(001) surface.

5. Summary

Photoelectron spectroscopy measurements of Cr(001) have revealed the periodicity of the near-surface valence-band dispersions along the [010] direction parallel to the crystal surface. The results indicate that the valence electrons feel, and self-consistently establish, antiferromagnetism in the Cr(001) near-surface layers.

Two surface valence electronic features are observed in normal-emission ARPES. The binding energy, symmetry, and dispersion of each feature is consistent with the latest theoretical calculations that predict a ferromagnetic Cr(001) surface phase on an otherwise antiferromagnetic Cr crystal. The temperature dependence of $E_{IN}(\underline{2})$ resembles a temperature-dependent ferromagnetic exchange splitting (Figure 5). This behavior is interpreted as evidence for a surface magnetic phase transition on Cr(001).

An ARPES study of the Cr(001) 3s core level revealed that the relative intensities of the exchange-split photoelectron components change as the ARPES spectra become more surface sensitive. Surface contamination suppresses this intensity variation and decreases the energy width of the 3s spectral profile. Almost all of the spectral observations are consistent with a surface-sensitive enhancement of the atomic 3d spin at the Cr(001) surface.

Acknowledgments.

This work was supported by the Diractor, Office of Energy Research, Office of Basic Energy Sciences, Chemical Sciences Division of the U. S. Department of Energy, under Contract No. DE-AC0376SF00098.

References.

1. L. E. Klebanoff, S. W. Robey, G. Liu, and D. A. Shirley, Phys. Rev. B 30, 1048 (1984).
2. L. E. Klebanoff, R. H. Victora, L. M. Falicov, and D. A. Shirley, Phys. Rev. B 32, 1997 (1985).
3. L. E. Klebanoff and D. A. Shirley, Phys. Rev. B 33, 5301 (1986).
4. L. E. Klebanoff, S. W. Robey, G. Liu, and D. A. Shirley, Phys. Rev. B 31, 6379 (1985).
5. Y. Teraoka and J. Kanamori, Transition Metals, 1977, ed. by M. J. G. Lee, J. M. Perz, and E. Fawcett, (The Institute of Physics, Bristol and London 1978) p. 588.
6. G. Allan, Surf. Sci. Rpts. 1, 121 (1981).

7. G. Allan, Phys. Rev. B 19, 4774 (1979).
8. G. Allan, Surf. Sci. 74, 79 (1978).
9. D. R. Grempel, Phys. Rev. B 24, 3928 (1981).
10. R. H. Victora and L. M. Falicov, Phys. Rev. B 31, 7335 (1985).
11. C. L. Fu and A. J. Freeman, Phys. Rev. B 33, 1755 (1986).
12. G. Shirane and W. J. Takei, J. Phys. Soc. Japan Suppl: 17 B III, 35 (1962).
13. A. W. Overhauser, Phys. Rev. 128, 1437 (1962).
14. N. I. Kulikov and E. T. Kulatov, J. Phys. F. 12, 2291 (1982).
15. H. L. Skriver, J. Phys. F: Met. Phys. 11, 97 (1981).
16. J. Kuebler, J. Magn. Magn. Mat. 20, 277 (1980).
17. S. Asano and J. Yamashita, J. Phys. Soc. Jpn. 23, 714 (1967).
18. H. Hasegawa, submitted to J. Phys. F.
19. Saturation coverage is achieved at 1L exposure; A. G. Baca, private communication.
20. D. E. Eastman, F. J. Himpsel, and J. A. Knapp, Phys. Rev. Lett. 40, 1514 (1978).
21. C. S. Fadley, D. A. Shirley, A. J. Freeman, P. S. Bagus, and J. V. Mallow, Phys. Rev. Lett. 23, 1397 (1969).
22. D. A. Shirley, in Topics in Applied Physics, Photoemission in Solids I, edited by M. Cardona and L. Ley (Springer-Verlag, Berlin, 1978), p. 165.
23. C. S. Fadley, in Electron Spectroscopy, edited by D. A. Shirley (North-Holland, 1972), p. 781.
24. S. P. Kowalczyk, Ph.D. Thesis, University of California, Berkeley, 1976.
25. F. R. McFeely, S. P. Kowalczyk, L. Ley, and D. A. Shirley, Solid State Commun. 15, 1051 (1974).
26. C. S. Fadley and D. A. Shirley, Phys. Rev. A 2, 1109 (1970).
27. J. H. Van Vleck, Phys. Rev. 45, 405 (1934).
28. P. A. Cox and F. A. Orchard, Chem. Phys. Lett. 7, 273 (1970).
29. P. A. Cox, J. Elect. Spect. Relat. Phenom. 22, 77 (1981).
30. P. S. Bagus, A. J. Freeman, and F. Sasaki, Phys. Rev. Lett. 30, 850 (1973).
31. J. R. Leite, J. C. Rodrigues, A. C. Ferraz, and A. C. Pavão, Phys. Rev. B 16, 978 (1977).
32. C. J. Powell, Surf. Sci. 44, 29 (1974).

Magnetic Resonance on Monolayers in Ultra High Vacuum

K. Baberschke, M. Zomack, and M. Farle

Institut für Atom- und Festkörperphysik, Freie Universität Berlin, Arnimallee 14, D-1000 Berlin 33, Germany

High sensitivity ESR on 1 to 10^{-2} monolayer - 10^{14} to 10^{12} spins - at ultra-high-vacuum clean surfaces is reported. A conventional ESR spectrometer for 9, 4, and 1 GHz is combined with an UHV chamber. The chamber is equipped with conventional apparatus for surface analysis as ion bombardment, low-energy electron diffraction, and Auger electron spectroscopy. This allows a proper adsorbate/surface preparation and analysis. We report: (i) The ferromagnetic resonance (FMR) below the bulk Curie temperature and the EPR above T_{cb} = 292 K of a monolayer of Gd(0001) on W(110). The Curie-Weiss temperatures of the mono- and doublelayer are <275 K and 280 K, respectively. (ii) The ESR of 10^{-2} Langmuir NO_2 molecules physisorbed on Kr/Ag(110) is detected. The symmetry axis of the NO_2 molecule lies in the surface plane.

1. Introduction.

Electron spin resonance has been used in various fields to investigate magnetic properties in the bulk solid, the liquid, and the gas phases. It is well established that the ESR is able to detect 10^{10} to 10^{11} spins for systems with narrow lines and under certain experimental conditions. For systems with a linewidth $\Delta H \approx 10$ to 100 G, caused by stronger spin-flip rates, 10^{13} to 10^{14} particles may be needed for detection. But this is still less than a monolayer of atoms on a 1 cm^2-surface (for details see [1]). Consequently, ESR seems to be a highly sensitive technique for the investigation of surface magnetism. For metal and semiconductor surfaces the surface analysis techniques can be separated into two groups: those which are sensitive to the first few atomic surface layers only and those which are not surface sensitive by themselves; for the latter the adsorbate surface has to be prepared independently. Techniques of the first type are, for example: electron-capture spectroscopy (ECS) [2], spin-polarized electron studies [3-5]; examples of the second type are: magnetization measurements [5,6], fluorescence photon detection [5], and the ESR reported here (the skin-depth of microwaves in metals is typically 1 µm; the escape-depth of fluorescence photons, depending on energy, is in the order of 1000 Å).

ESR has investigated the catalytic reactions of oxygen on surfaces [7]. Because of technical factors the surfaces are, in most cases, not well defined and are investigated in technical vacuum only (~10^{-6} mbar). The Cornell group [8] has extended this work to UHV (~10^{-10} mbar), but they concentrated more on gas-surface reactions at high pressure (~1 mbar). For semiconductor surfaces cleaved in UHV an ESR signal caused by to dangling bonds (in micro- cracks) was detected [9]. However, this signal was not found on clean and perfect surfaces [10]. Using the Langmuir-Blodgett method, 2D manganese magnets were prepared and measured by ESR [11]. The angular dependence of the linewidth shows a 2D-behavior. The work which seems to be very close to the present one is the recent "Conduction electron spin resonance on Li-films in UHV" by S. Schultz and coworkers [12]. They prepare and analyze the Li-film in UHV and study the change of ΔH as a function of adsorption of various gases and metals on the surface.

In the present work we report the first ESR of local moments of submonolayer coverage of (i) NO_2 molecules on Ag(110) and (ii) of Gd adsorbed on W(110). As far

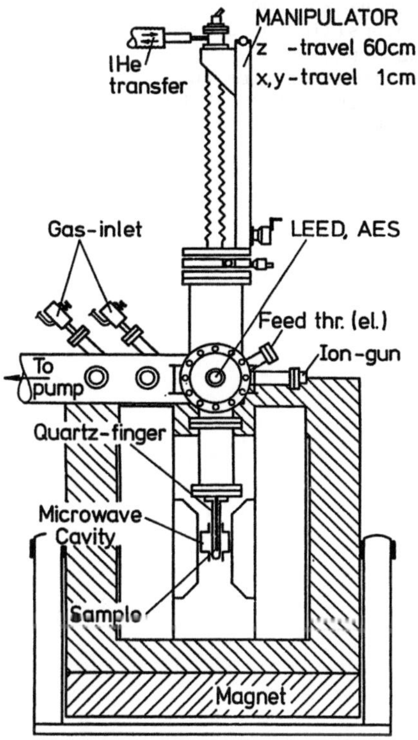

Figure 1. Experimental set-up of an UHV chamber ($p < 2x10^{-10}$ mbar) and a commercial ESR spectrometer. The sample is in ultra high vacuum, the microwave cavity in normal laboratory air.

as we are aware these are the first experiments where a full in situ surface analysis is available, single crystals are used and are rotatable with respect to H_o, and the sample temperature can be varied between 400 K > T > 15 K. In Section 2 we describe the experimental set-up. Section 3 presents the results of a monolayer of Gd(0001)/W(110). This is of particular interest with respect to recent SPLEED measurements [13]. We have chosen the same system and did follow the same sample preparation as the Jülich group, getting the same AES results as [14]. Finally, in Section 4 the ESR of a 10^{-2} layer of NO_2-molecules on an inert gas-metal substrate - Kr/Ag - is analyzed.

2. Experimental Set-Up of ESR in UHV.

There exist two principal possibilities to combine an ESR spectrometer with an UHV chamber: (i) The microwave cavity is installed inside the UHV chamber [8], or (ii) the microwave cavity is kept in normal laboratory air and an UHV clean quartz finger tip is inserted into the cavity. Our group has experience in the latter case from ESR in the millikelvin regime [15]. Figure 1 shows the apparatus. Shortly we will describe its function and why we find it to be superior to other systems. The quartz finger tip has an outer diameter of 25 mm and fits into commercial high-Q cavities. We are in a position to exchange a 9 GHz cavity with 4, 3, or 1 GHz within minutes. The long z-travel of the manipulator allows to move the specimen down into the ESR position or up into the sample preparation level. In the upper position all necessary surface analytical techniques may be installed. The present chamber includes: Low Energy Electron Diffraction (LEED), Auger Electron Spectroscopy (AES) with a single pass CMA, sputter gun, gas dosing, and evaporation system etc. The base pressure is $<2x10^{-10}$ mbar at 300 K. A liquid He cooling system installed in the manipulator allows to cool the sample down to ~15 K. During low-temperature experiments the base pressure is better than $2x10^{-11}$ mbar. Some preliminary ESR results are reported in [1].

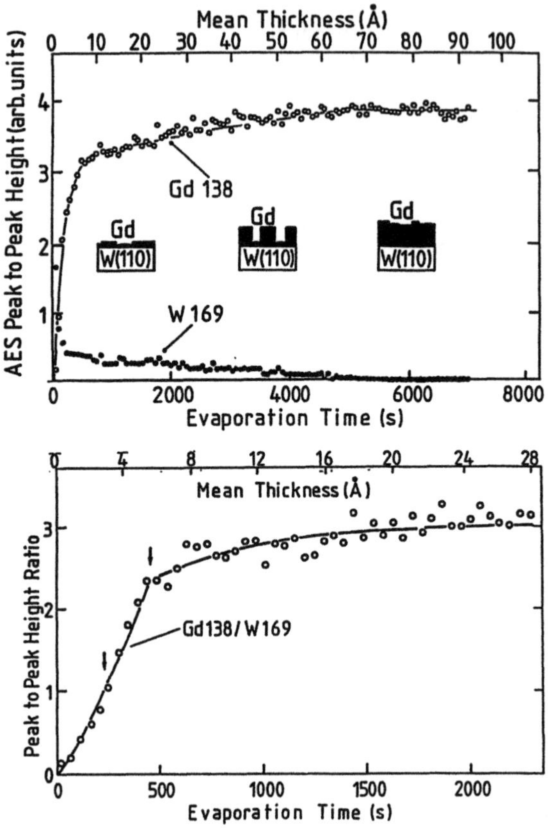

Figure 2. Auger intensities of Gd and W as function of evaporation time and mean thickness. Taken from [14].

3. ESR and FMR of a Monolayer Gd on W(110).

Following the early work by C. RAU [2], the Jülich group investigated extensively the magnetic ordering of the surface layer of Gd on an approximately 100 Å thick Gd film/W(110) by spin polarized LEED (SPLEED). In Figure 2 (taken from [14]) we summarize their results on the growth of such a film. There are three steps of adsorption: (i) at small evaporation rate a monolayer with regular hexagonal structure and 3.9 Å Gd-Gd distance is formed (first arrow, mean thickness ~3.5 Å), further evaporation forms a second B-layer on top of the first hcp-A-layer (second arrow, mean thickness ~7 Å). (ii) At higher evaporation rates Stranski-Krastanow growth starts. (iii) Only at about 100 Å mean thickness a multilayer film with a smooth surface is formed. The last case may be discussed as a semi-infinite system and was measured by SPLEED [3,13].

In the present work we focus on step (i). This seems to be an almost ideal case of a 2D system. In Figure 3 experimental recorder traces are shown. At 297 K a paramagnetic resonance signal (Dysonian profile because of coupling to the conduction electrons) is recorded at g ≈ 2. At 276 K a collective excitation of spin waves (FMR) is seen; the intensity increases in the vicinity of T_c (T_{cb} = 292 K) very strongly because of the enhanced Curie-Weiss susceptibility. The shifts of the field for resonance and of the integrated intensity $I_{pp} \times \Delta H^2 \propto \chi$ are plotted in Figure 4 as a function of T for the monolayer (open circles, full circles) and for the double layer (open triangles, full triangles). In both cases the coordination number changes, namely Z = 6 and 9. The coupling strength J_{ij} may be different too because of RKKY oscillations. The inflection point of the intensity curves in Figure 4 can be seen as a magnetic ordering temperature. The

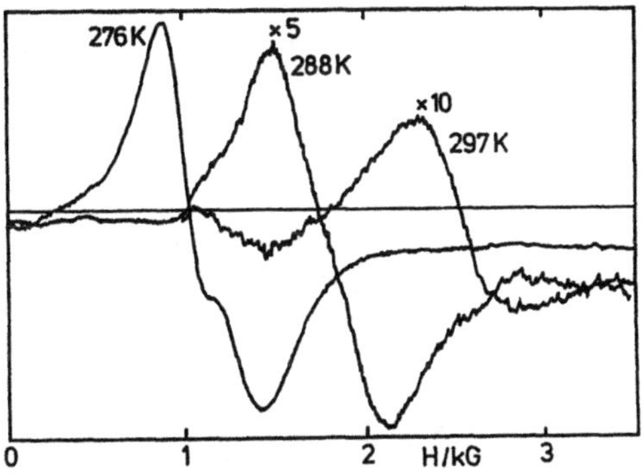

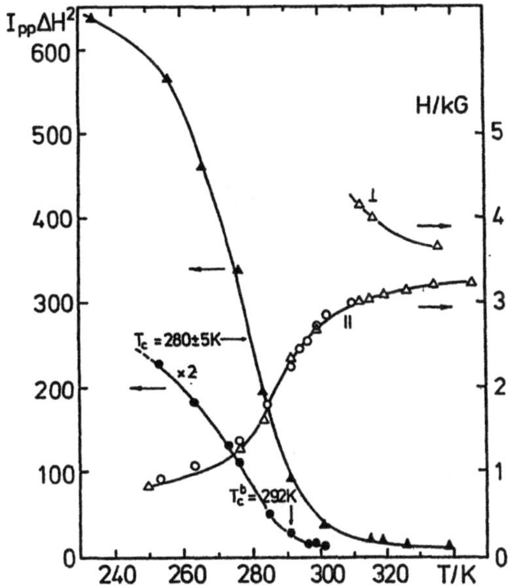

▲Figure 3. Magnetic resonance of a doublelayer (7 Å) of Gd(0001)/W(110). The ESR signal at 297 K was recorded with a 10x larger gain than the FMR signal at 276 K.

Figure 4. Resonance field (open symbols) and integrated intensities (full symbols) of a mono - (circles) and a double - (triangles) layer (see Figure 2). ▶

area under the ESR absorption signal is proportional to the static susceptibility. We have plotted $(I_{pp} \times \Delta H^2)^{-1}$ as a function of T. All data from high T down to the inflection point follow nicely a Curie-Weiss law with T_{cs} = 281 K, in agreement with Figure 4. The data for the monolayer are less complete, but one may find $T_{cs} \leq 275$ K.

A detailed analysis of this work will be given elsewhere [17]. For the present we conclude: a mono- and a double layer of Gd on W(110) do not show an enhanced magnetic ordering temperature. The Curie-Weiss temperatures are <275 K, ≈281 K, and 292 K (bulk). For an isotropic 2D Heisenberg system like Gd, the Mervin-Wagner theorem predicts no magnetic ordering for $T \neq 0$. But small anisotropic energies as dipolar interactions and single ion crystal field effects (<1 K) are always present and allow magnetic ordering.

4. 2D Arrangement of NO_2 on a Krypton/Ag(110) Substrate.

Previously [1] we have discussed why the spin flip rate of magnetic molecules on a clean metal substrate may be too high to be detectable by ESR. Consequently, we have reduced the coupling strength by bringing one to three layers of inert gas (Xe or Kr) between NO_2 and the Ag(110) surface. Figure 5 shows the resonance of 10^{-2} of a layer for an applied field H_0 in the surface plane. The figure demonstrates the high sensitivity as discussed in the introduction and - even more important - shows how ESR can contribute to solve questions of chemi- or physisorption. After adsorption of some gas on clean metal surfaces and conventional analyses [AES, LEED, PES, thermal desorption (TDS), etc.], usually many possible processes have to be considered: adsorption of the full molecule, dimerization, and fractional dissociation for instance must be discussed. Photoelectron spectroscopy is very sensitive to the valence state of an element, e.g. oxygen. But it does not so clearly distinguish whether a PE signal from oxygen comes from NO_2, N_2O_4, or NO molecules. In contrast, ESR will give no signal for the dimer N_2O_4 and distinguishes clearly between NO_2 and NO. The "fingerprint" (g-tensor and hyperfine coupling of S = 1/2 to the ^{14}N nucleus) of NO_2 is so clear and known from gas phase ESR, that a misinterpretation is impossible. We have measured the ESR from 10^{-2} to 100 L of NO_2, for different thicknesses of Kr, and at various temperatures. The full analysis will appear soon [18]. Here we restrict ourselves on the demonstration of the high sensitivity and the interpretation of the spectrum

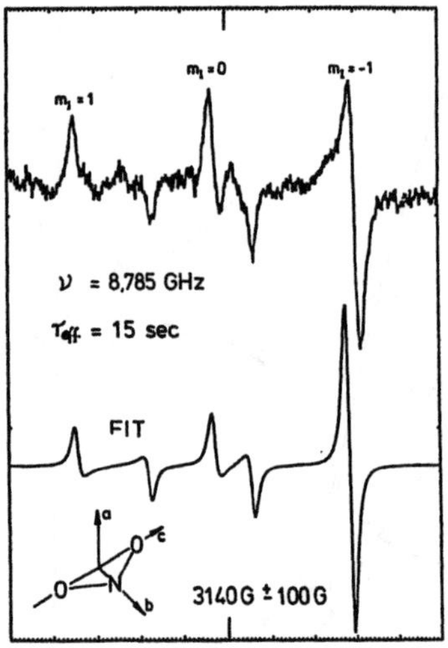

Figure 5. ESR at 20 K of 10^{-2} monolayer of NO_2 on three layers of Kr physisorbed on a Ag(110) single crystal. H is in the surface plane. The simulation (lower part) is identical to Figure 6(c).

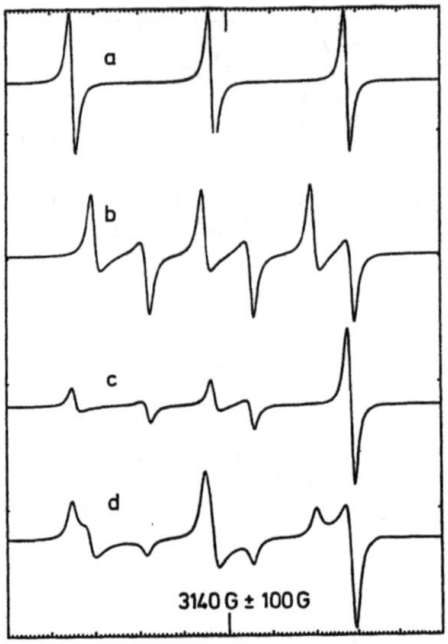

Figure 6. 4 simulations for different NO_2 orientations on the surface: a) b || H; b) b ⊥ H; c) a ⊥ H; d) a,b,c random.

in Figure 5. ^{14}N has a nuclear spin of I = 1; this splits the spectrum in Figure 5 into $m_I = \pm 1,0$ groups. Each group (i.e. $m_I = 0$) separates at most in three lines depending on the orientation of the a-, b-, and c-axis with respect to H. (For the free molecule $g_a = 2.006178$, $g_b = 2.00199$, and $g_c = 1.991015$.) In Figure 6 we simulated several models; input parameters are the g- and A-tensor of the molecule and a linewidth ΔH. In (a) we assumed all molecules with the a-axis (see insert Figure 5) parallel to the applied field, in (d) a random orientation of the a-, b-, and c-axis with respect to H is assumed. None of these two fits the experimental result. From arguments of chemisorption it is more likely to assume a 2D arrangement. In Figure 6(c) the molecule lies flat on the surface (c and b random), and in 6(b) the b-axis is ⊥ to H (in plane) but a and c have random distributions with respect to H_o. Figure 6(c) fits the experiment perfectly and is replotted in the lower part of Figure 5. For the sake of simplicity we have discussed a purely statistical (T=0) picture. More strictly speaking, the experiment proves only that the symmetry axis of the molecule |b| is in the surface plane. One cannot exclude a rotation along the b-axis. But more detailed data analysis makes this very unlikely [18].

In summary: We have demonstrated a new application of ESR in surface physics. An UHV chamber with in situ LEED, AES, etc. as well as liquid He cooling has been combined with angular dependent, high-sensitivity ESR.

Acknowledgments

We would like to thank S. F. Alvarado for several discussions and for making [14] available to us prior to publication. The work is supported by Deutsche Forschungsgemeinschaft, Sfb-6.

References

1. M. Farle, M. Zomack, and K. Baberschke, Surf. Sci. 160, 205 (1985).
2. Ç. Rau, this volume.
3. S. F. Alvarado, this volume.

4. D. T. Pierce, this volume.
5. U. Gradmann, this volume.
6. M. Dresselhaus, this volume.
7. J. H. Lunsford, Catalysis Rev. 8, 135 (1973), and Adv. Catalysis 32, 265 (1972).
8. M. Nilges, M. Shiotani, C. T. Yu, G. Barkley, Y. Kera, and J. H. Freed, J. Chem. Phys. 73, 588 (1980), and M. Nilges, G. Barkley, M. Shiotani, and J. H. Freed, J. de Chémie Physique 78, 910 (1981).
9. B. P. Lemke and D. Hanemann, Phys. Rev. B 17, 1893 (1978).
10. We have recently repeated this experiment on Si(111) single crystal with a 7x7 overstructure. No ESR signal was detectable, in agreement with the recent literature.
11. H. Pomerantz, Surf. Sci. 142, 556 (1984).
12. D. M. Eigler and S. Schultz, Solid State Comm. 44, 1565 (1982), and D. M. Eigler and S. Schultz, Phys. Rev. Lett. 54, 1185 (1985).
13. D. Weller, S. F. Alvarado, W. Gudat, K. Schröder, and M. Campagna, Phys. Rev. Lett. 54, 1555 (1985).
14. D. Weller and S. F. Alvarado, J. Appl. Phys., to appear (1986).
15. Y. V. Spalden, E. Tsang, K. Baberschke, and P. Schlottmann, Phys. Rev. B 28, 24 (1983) and references therein.
16. C. Rau, J. Magn. Magn. Mat. 30, 141 (1982).
17. M. Farle and K. Baberschke, to be published.
18. M. Zomack and K. Baberschke, ECOSS 8, Jülich, April 1986, to appear in Surf. Sci.

New Applications of Electron Capture Spectroscopy

C. Rau

Physics Department, Rice University, Houston, TX 77251, USA

Electron capture spectroscopy (ECS) has been used to probe the electron-spin polarization (ESP) at the topmost atomic layer at Ni(111), fcc γ-Fe(111), V(100), and Tb surfaces. Contrary to bulk behavior, for the Ni(111) surfaces the ESP-signal, which measures the long-range surface magnetic order, decreases from -43% at T = 293 K almost linearly to zero at the bulk Curie temperature T_{Cb} = 631 K. "Local" surface magnetic order is temperature independent and is detected far above T_{Cb}. Such a behavior is only found for the transition metal Ni. For surfaces of the 4f-metal Tb, the "local" surface magnetic order is strongly temperature dependent and disappears above T_{Cb}. For (111)-surfaces of ultrathin epitaxial layers of fcc γ-Fe, pseudomorphically deposited on Cu(111) substrate crystals, two-dimensional ferromagnetic order is detected at room temperature. For the topmost atomic layer of magnetized (100)-surfaces of bulk paramagnetic V, long-range ferromagnetic order is found. The ESP decreases linearly from -30% at room temperature to zero at a surface Curie temperature T_{Cs} = 540 K.

1. Introduction.

In the past ten years, surface and thin-film magnetism has become a field of increasing scientific and technological interest [1,2]. Recent theoretical advances in developing accurate and reliable methods for the calculation of spin-polarized surface and thin-film band structures [3-5] and the development and implementation of powerful experimental tools (photo-electron emission [2,6], electron capture [7], scattering of spin-polarized electrons [6,8] and metastable He atoms [9,10], Mössbauer spectroscopy [11], magneto-metric [11] and magnetooptic [12] measurements), for the investigation of surface magnetic order, now open a fascinating and promising way to answer many fundamental questions about surface magnetism. Can long-range or local ferromagnetic order exist at surfaces of bulk antiferro- or paramagnetic materials? What changes at the critical temperatures? Theoretical and experimental data on the surface magnetic order, which is well characterized by the electron spin polarization [2,10], and its temperature dependence at the surface of bulk ferro-, antiferro-, or paramagnetic materials can give a direct answer to these challenging questions.

We report on experimental results on the long-range and local magnetic order at surfaces of Ni(111), fcc γ-Fe(111), V(100), and Tb utilizing ECS, which is extremely sensitive to the magnetic order existing at the topmost atomic layer of a surface.

ECS allows us to investigate long-range or macroscopic magnetic order and also local or microscopic magnetic order existing on an atomic scale over a few atomic spacings along a surface [13]. In addition, ECS allows us to measure the ESP as a function of the applied external magnetic field, which is very important for the investigation of surface critical behavior.

For Ni(111) surfaces, it is found that the long-range ESP decreases almost linearly with temperature and disappears around the bulk Curie temperature T_{Cb}. Local magnetic order, however, is still detected far above T_{Cb}. At (111) surfaces of ultrathin, epitaxial layers of fcc γ-Fe, pseudomorphically deposited on Cu(111) substrate crystals, long-range and local ferromagnetic order is detected. For magnetized bulk paramagnetic V, it is found that long-range and local ferromagnetic order exists at the topmost atomic layer of the (100)surface. The long-range ESP decreases linearly, from around -30% at room temperature, to zero at a surface Curie temperature T_{Cs} = 540 K. Contrary to surfaces of the 3d-transition metal Ni, at surfaces of 4f-rare-earth metal Tb, the local magnetic order disappears above the bulk critical temperature.

2. Experimental.

The basic process in ECS [14,15] is the capture of one or two electrons during grazing-incidence reflection of fast deuterons at magnetic surfaces. For an angle of incidence α = 0.2°, the ions are specularly reflected at the surface. The distance of closest approach d_{min} of ions of energy 150 keV is ~2 A [7]. Thus, the ions probe in real space only the exponential tails of the electronic wave functions at the surface, which gives the extreme surface sensitivity of ECS.

Figure 1 gives a schematic representation of the experimental set-up used in ECS. A well-collimated deuteron beam is incident at grazing angle on the magnetic target, which is magnetized parallel to the surface plane. After reflection, the beam passes through a transverse electric field, which spatially separates remaining D^+ from D^0 atoms and D^- ions formed by one- and two-electron capture, respectively. The ion currents are measured using two Faraday cups. The undeflected D^0 atoms, however, impinge on a Ti-tritium target where, by measuring the asymmetry in the angular distribution of the α-particles emitted in the reaction $T(d,n)\alpha$, their nuclear polarization is determined.

Long-range surface magnetic order is detected using one-electron capture processes. At a ferromagnetic surface, one-electron capture results in the formation of D^0 atoms polarized in the electron shell. This ESP is transferred by hyperfine interaction to the deuterium nucleus. The nuclear polarization is a direct measure of the spin polarization of the captured electrons. The electron-spin polarization, P, along the target magnetizing field is defined by $P = (n^+ - n^-)/(n^+ + n^-)$, where n^+ and n^+ represent the fractional numbers of electrons with spin moment antiparallel (majority electrons) and parallel (minority electrons), respectively, to the target magnetizing field. Electrons captured by different deuterons originate at widely separated points on the magnetic surface. Therefore, long-range surface magnetic order is detected with one-electron capture processes.

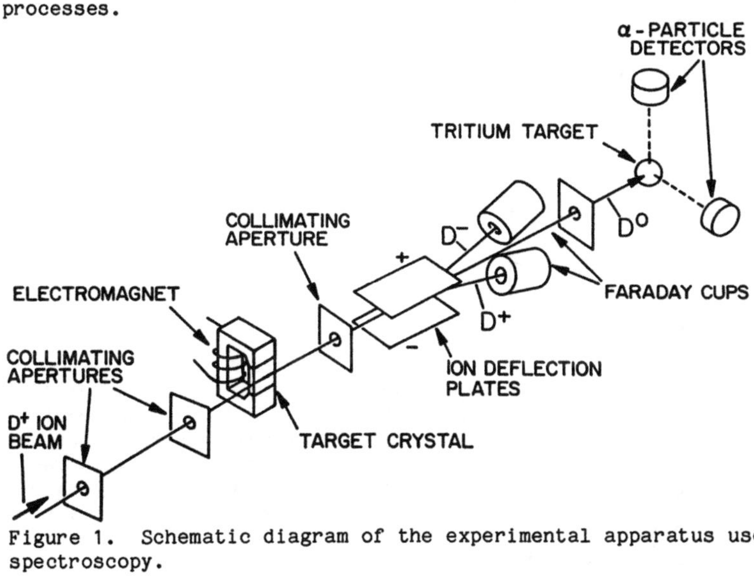

Figure 1. Schematic diagram of the experimental apparatus used in electron capture spectroscopy.

In addition, ECS allows us to investigate local ferromagnetic order (spin-spin correlations) on an atomic scale. We utilize two-electron capture processes, which yield negative deuterium ions. Here the situation is significantly different. The only stable state of D^- is the $1s^2\ ^1S$ state. Therefore, D^- ions can only be formed by capture of electrons of antiparallel spin orientations. The characteristic length within an incident deuteron captures two electrons is small, ~10-20 Å, for 150 keV deuterons. Thus, two captured electrons originate in the same "local" surface region, and the two-electron capture will be strongly suppressed by the existence of "local" ferromagnetic order at a surface. The reduction in the D^- yield in the reflected beam, relative to that observed for a nonmagnetic sample such as Cu, provides a direct measure of the local surface magnetic order existing at a surface on an atomic scale. For further details we refer to [1,13].

3. Results and Discussion.

3.1 Temperature Dependence of Ferromagnetism at Ni(111) Surfaces.

At present there is, contrary to a wealth of theoretical information on surface phase transitions [16], a lack in experimental information on the temperature dependence of the long-range and local magnetic orders at surfaces of bulk and thin film magnetic materials. Highly desirable are measurements over a wide range of temperatures and also very close to the critical temperature, which is defined by the disappearance of the long-range magnetic order. One of the reasons for the lack of experimental data is that there are not many experimental techniques which allow reliable data to be obtained over a wide range of temperatures. Furthermore, intrinsic surface effects can only be detected if the experimental probing depth is of the order of, or less than, a lattice constant normal to the surface.

We have already successfully shown that for atomically clean and flat surfaces of the 4f-rare-earth metal Gd, the temperature dependence of the long-range surface magnetic order is drastically different from the bulk behavior [17,18]. Surface Curie temperatures were found to be far above the bulk Curie temperature.

We have measured the surface magnetic order at surfaces of the 3d-transition metal Ni between 293 K and 940 K (see upper part of Figure 2). By means of one-electron capture processes, we find that, contrary to the T-dependence of the

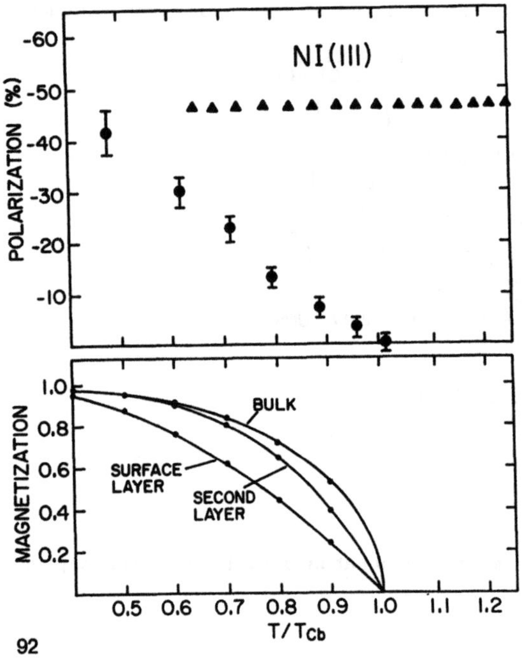

Figure 2. Upper part. Temperature dependence of the ESP signal originating from long-range (circles) and local (triangles) surface magnetic order on Ni(111). Lower part. Temperature dependence of the long-range magnetization of the first and second layer and of the bulk in mean-field theory [19,20].

bulk behavior (see lower part of Figure 2), the long-range surface magnetic order decreases linearly with increasing temperature until it drops to zero at the bulk Curie temperature T_{Cb} and remains zero above T_{Cb}. The measurements are performed on atomically clean and flat surfaces of Ni(111). At all temperatures, C, S, and O coverages were less than 0.02 monolayer, as monitored with Auger electron spectroscopy.

The lower part of Figure 2 shows a plot of the T-dependence of the magnetization of layers parallel to the (111) surface of a fcc ferromagnet in the mean-field model [19-21]. The ESP signal caused by long-range magnetic order, measured at the topmost surface layer, follows closely the temperature behavior of the magnetization of that layer, as predicted by mean-field theory, This theory is expected to be qualitatively correct [21] when the spin-spin correlation length is not large compared to the lattice spacing (valid for $T/T_{Cb} < 0.9$).

The temperature dependence of the local surface magnetic order at Ni(111) surfaces shows significantly different behavior. The local ESP [13] [the triangles in Figure 2) remains constant until the highest measured temperature of 940 K, which is far above T_{Cb}, implying that exchange effects between minority- and majority-spin electrons from nearby atoms are temperature independent [13]. We remark that this behavior of the local surface magnetic order is only found for the 3d transition metal Ni.

ECS measurements very close to T_{Cb} are underway to investigate surface critical exponents. These include measurements on the magnetic-field dependence of the ESP in the critical region, which is important for testing scaling laws in surface magnetism [22].

A particular advantage of ECS, in comparison with other spin-sensitive spectroscopies -- where electrons are used, and therefore even weak magnetic fields influence the electron trajectories -- is that the high-energy ions used in ECS are nearly unaffected even by rather large magnetic fields.

3.2 Ferromagnetic Order at Surfaces of Ultrathin, Epitaxial fcc γ-Fe(111)p(1x1) Films on Cu(111).

The high temperature phase, fcc γ-Fe, of bulk Fe is usually not stable at room temperature. It can, however, be stabilized by deposition of ultrathin, epitaxial Fe films on surfaces of fcc Cu substrates [23,24].

In a series of experimental investigations (magnetometric measurements, neutron diffraction, Mössbauer investigations, susceptibility measurements, etc.) controversial experimental results were obtained. These films were either found to be ferromagnetic at room temperature [25] or paramagnetic at 300 K showing a transition to antiferromagnetism at lower temperatures [26]. For further detailed information the reader should see Reference [27].

Under ultra-high vacuum conditions and by means of electron-beam evaporation, we have prepared well-defined, atomically clean and flat fcc γ-Fe(111) films (1-4 layers) on Cu(111) substrate crystals. During evaporation, the substrates were kept at 373 K and the evaporation rate was kept between 0.01 and 0.3 Å/s, where it is well known [24] that below a layer thickness of five layers fcc γ-Fe(111) on Cu(111) no α-phase transition occurs. These layers, therefore, can be pseudomorphically grown on Cu substrate crystals. The absence of holes in these films is checked by monitoring Auger electron signals, and the monocrystalline fcc state is detected by low-energy electron diffraction [27].

The magnetic order at the surface of these layers is measured using ECS. For four layers of fcc γ-Fe(111)p(1x1) on Cu(111), the ESP signal, measuring long-range magnetic order, amounts to +11% at 373 K and to +16% at 298 K. This clearly proves, for this temperature region, the existence of ferromagnetic order at the surface of these films. The presence of local ferromagnetic order at surfaces of

1-4 monolayers fcc γ-Fe(111)p(1x1) on Cu(111) is detected from the ESP signal due to two-electron capture processes. We find nonzero local ESP between 300 K and 400 K.

Our measurements, performed on free (uncovered) surfaces, are consistent with GRADMANN's magnetometric measurements [24] on fcc Cu(111)/fcc γ-Fe(111)/ Cu(111) sandwich films. Experiments by GONSER and HALBAUER [26], utilizing Mössbauer spectroscopy for the investigation of (111)-oriented CuFeCu sandwich films, revealed the existence of antiferromagnetic order below 20 K-40 K and paramagnetic order at room temperature. The authors note in their paper that all films (thickness between 18-25 Å) revealed an island structure. We believe that the discrepancies in the experimental results between the various methods are possibly due to the application of different sample preparation techniques.

3.3 Magnetic Order at (100) Surfaces of Bulk Paramagnetic Vanadium.

Solid vanadium possesses a high paramagnetic susceptibility, but does not exhibit ferromagnetism in the bulk. Isolated vanadium atoms possess a permanent magnetic moment of 3 μ_B in the ground state. In the intermediate situation, e.g., for vanadium surface atoms, where the number of nearest neighbors is reduced, one might still expect magnetic behavior. In 1979, ALLAN [28] showed that, due to magnetic instabilities localized near the surface, a permanent magnetic moment might exist at the V(100) surface. YOKOYAMA and coworkers [29] reached similar conclusions for the (100) surface of a five-layer thick V film. Recently, FREEMAN and coworkers [3] predicted ferromagnetic behavior for one monolayer of vanadium on Ag(100).

For a first set of ECS experiments at V surfaces, we selected the (100) surface where a surface atom has only four nearest neighbors compared to eight in the bulk and investigated the long-range and local magnetic order between 300 K and 650 K. With the application of a magnetizing field H = 41 kA/m, we find at 300 K an ESP of up to -30%, a fact which clearly demonstrates the existence of long-range ferromagnetic order at the topmost atomic layer of V(100). The ESP decreases almost linearly and drops to zero at a surface Curie temperature T_{Cs} = 540 K.

In these experiments the residual oxygen contamination at the surface was less than 2%. We believe that the nonzero ESP cannot be caused by the existence of vanadium oxides which, to our knowledge, possess Néel temperatures below 400 K, whereas in our ECS measurements ferromagnetic order is detected up to 540 K. We note that all our ECS experiments where we investigated the influence of an oxygen contamination on the surface magnetic order (Ni, Co, Fe, Gd, Tb), without an exception, revealed a reduction of the original ESP by residual oxygen at the surface. The intrinsic (higher) polarization value of the clean surface always could be restored by cleaning the oxygen contamination from surfaces. Therefore, our present data yield possibly a lower bound for the value for the magnitude of the intrinsic ESP at the oxygen free, magnetized (H = 41 kA/m) V(100) surface.

3.4 Temperature Dependence of the Local Magnetic Order at Tb Surfaces.

As reported in Section 3.1, our recent ECS experiments [13] at Ni(hkl) surfaces show that local surface magnetic order exists at an atomic scale, within a few atomic neighbors, far above the Curie temperature of Ni, which is the prototype of an itinerant-electron ferromagnet. For localized 4f-electron Heisenberg ferromagnets such as Tb, one expects different behavior. Contrary to the 3d transition metals, where the electrons responsible for ferromagnetism are itinerant and located within a few eV of the Fermi level, in the 4f-rare-earth metals the magnetic electrons are localized in real space. The interatomic alignment of the 4f-electrons is produced by an indirect interaction via polarized conduction electrons (RKKY model) which -- contrary to the 4f-electrons -- are available for capture at the surface. Above the critical temperature, where the transition to paramagnetic behavior occurs, the 4f-electron polarization disappears on an atomic scale between atomic neighbors. This should also result in a disappearance of the local conduction electron polarization.

Terbium is ferromagnetic below T_{Cb} = 220 K, antiferromagnetic with a helical spin structure up to its Néel temperature T_{Nb} = 228 K, and paramagnetic above this temperature [30]. By means of ECS we find, for Tb surfaces at 220 K, a local ESP of 68% which decreases to 56% at 250 K and to 15% at 270 K, and disappears at around 300 K. By decreasing the temperature again to 220 K, the previously measured polarization values can be restored.

Such a temperature dependence of the local ESP at Tb surfaces is drastically different from that found at Ni surfaces. In our next ECS experiments on Tb surfaces, we are planning to vary the velocity of the deuterons which allows us to change the mean distance, ℓ, within which two electrons are captured. This should give a deeper insight into the ℓ-dependence of spin-spin correlations at magnetic surfaces. Experiments to investigate the temperature dependence of the long-range surface magnetic order are also in progress in our group. They are expected to yield the exact value of T_{Cs}.

Acknowledgment.

This work was supported by U. S. National Science Foundation under Grant No. DMR-8406975.

1. C. Rau, J. Magn. Magn. Mat. 30, 141 (1982).
2. H. C. Siegmann, F. Meier, M. Erbudak, and M. Landolt, Adv. El. El. Phys. 62, 1 (1984).
3. C. L. Fu, A. J. Freeman, and T. Oguchi, Phys. Rev. Lett. 54, 2700 (1985).
4. R. H. Victora and L. M. Falicov, Phys. Rev. B 13, 7335 (1985).
5. R. Richter, J. G. Gay, and J. R. Smith, Phys. Rev. Lett. 54, 2704 (1985).
6. Polarized Electrons in Surface Physics, ed. by R. Feder (World Scientific, Singapore, 1985).
7. C. Rau, Comm. Solid State Phys. 9, 177 (1980).
8. D. T. Pierce and R. J. Celotta, Adv. Electron. Electron Phys. 56, 219 (1981).
9. M. Onellion, M. W. Hart, F. B. Dunning, and G. K. Walters, Phys. Rev. Lett. 52, 380 (1984).
10. F. B. Dunning, C. Rau, and G. K. Walters, Comm. Solid State Phys. 12, 17 (1985).
11. U. Gradmann, this volume.
12. S. D. Bader, E. R. Moog, and P. Grünberg, this volume, and J. Magn. Magn. Mat. (in press).
13. C. Rau and S. Eichner, Phys. Rev. Lett. 47, 939 (1981).
14. C. Rau and R. Sizmann, Phys. Lett. 43A, 317 (1973).
15. C. Rau and R. Sizmann, in Atomic Collisions in Solids, ed. by S. Datz et al. (Plenum, New York, 1975), Vol. 1, p. 295.
16. K. Binder, in Phase Transitions and Critical Phenomena, ed. by C. Domb, M. S. Green, and J. L. Lebowitz (Academic, New York, 1983), Vol. 8, and references therein.
17. C. Rau and S. Eichner, in Nuclear Phys. Methods in Mat. Research, ed. by K. Bethge et al. (Vieweg, Braunschweig, 1980), p. 354.
18. C. Rau, J. Magn. Magn. Mat. 31-34, 874 (1983).
19. T. Wolfram, R. E. DeWames, W. F. Hall, and P. W. Palmberg, Surf. Sci. 28, 45 (1971).
20. E. Kisker, J. Phys. Chem. 87, 3597 (1983).
21. S. W. Wang, Solid State Comm. 36, 847 (1980).
22. B. Widom, J. Chem. Phys. 43, 3898 (1965).
23. J. G. Wright, Phil. Mag. 24, 217 (1971).
24. U. Gradmann and P. Tillman, phys. stat. sol. (a) 44, 539 (1977).
25. W. Kümmerle and U. Gradmann, phys. stat. sol. (a) 45, 171 (1978).
26. R. Halbauer and U. Gonser, J. Magn. Magn. Mat. 35, 55 (1983).
27. C. Rau, C. Schneider, G. Xing, and K. Jamison, private communication and to be published.
28. G. Allan, this volume, and Phys. Rev. B 19, 4774 (1979).
29. G. Yokoyama, N. Hirashita, T. Oguchi, T. Kambara, and K. I. Gondaira, J. Phys. F 11, 1643 (1981).
30. R. D. Greenough and N. F. Hettiarachchi, J. Magn. Magn. Mat. 31-34, 178 (1983).

Part III

Statistical Mechanics of Surfaces and Interfaces

Influence of Dilution and Nature of the Interaction on Surface and Interface Magnetism

C. Tsallis

Permanent address: Centro Brasileiro de Pesquisas Físicas/CNPq, Rua Xavier Sigaud 150, 22290 Rio de Janeiro, Brazil and Departamento de Física Teórica e Experimental, Universidade Federal do Rio Grande do Norte, Cidade Universitária, 59000 Natal-RN, Brazil

The recent theoretical effort of the Rio de Janeiro/CBPF group on surface magnetism is tutorially reviewed. Within a real-space renormalization group framework, we analyze the influence of factors such as the number of states per spin (q-state Potts model), the signs of the coupling constants (mixed ferro- and antiferromagnetic interactions), the presence of a second semi-infinite bulk (interface case), the symmetry of the interaction (anisotropic Heisenberg model), and surface and/or bulk dilution (bond quenched model). A variety of interesting physical effects emerges.

1. Introduction.

During the last decade, surface magnetism has raised considerable interest both because of its various applications (catalysis, corrosion, etc.) and its intrinsic theoretical and experimental richness. Itinerant as well as localized-ion magnetic systems have been studied. Nevertheless, the field can be considered as being at its initial stage: this is due to real experimental and theoretical difficulties associated with it.

Surface magnetic order has been experimentally exhibited on systems such as Ni, Cr, Gd [1-8]. On theoretical grounds, the problem has been treated within different frameworks, namely mean-field approximation [9-11], various effective field theories [12-15], the Bethe approximation [16], series expansion [17], the random-phase approximation [18], Monte Carlo techniques [19], and renormalization-group (RG) methods [20-30]. Several among these works (as well as others) have been reviewed by BINDER [31]. We present here a comprehensive review of the theoretical effort [32-40] that has been very recently accomplished on the subject by the Rio de Janeiro/CBPF group and collaborators. We shall discuss, within the real space RG framework, the influence on the surface-magnetism criticality (phase diagrams and universality classes) of the following factors: (a) the number of states per spin (q-state Potts model; Section 3); (b) the signs of the coupling constants (mixed ferro- and antiferromagnetic interactions; Section 4); (c) the presence of a second semi-infinite bulk (interface case; Section 5); (d) the symmetry of the interaction (anisotropic Heisenberg model; Section 6); (e) surface and/or bulk dilution (bond quenched model; Section 7). To understand better the influence of all these factors, we shall first focus a prototype, which we choose to be the semi-infinite simple cubic spin 1/2 Ising ferromagnet with the (1,0,0) free surface.

2. Prototype.

The Hamiltonian of our prototype is given by

$$\mathcal{H} = -\sum_{\langle i,j\rangle} J_{ij}\,\sigma_i\,\sigma_j \qquad (\sigma_i = \pm 1) \quad , \tag{1}$$

with $J_{ij} = J_s > 0$ if both i and j sites belong to the (1,0,0) surface, and $J_{ij} = J_B > 0$ otherwise. It is convenient to define $\Delta \equiv J_S/J_B - 1$.

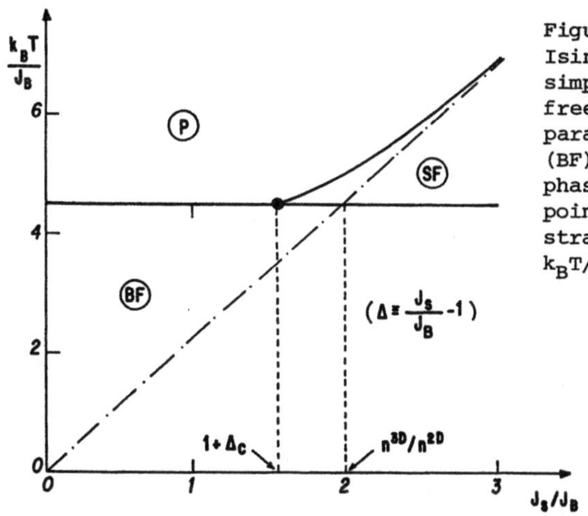

Figure 1. Phase diagram of the spin 1/2 Ising ferromagnet in the semi-infinite simple cubic lattice with the (1,0,0) free surface (prototype model). The paramagnetic (P), bulk ferromagnetic (BF), and surface ferromagnetic (SF) phases join at the SB multicritical point (full circle). The $\Delta \to \infty$ asymptotic straight line (dot-dashed) satisfies $k_BT/J_B = n^{2D}J_S/J_B$.

The phase diagram of this system is known to be as given in Figure 1. For $T < T_c^{3D} \equiv n^{3D}J_B/k_B$ (with $n^{3D} \approx 4.511$) we have the bulk ferromagnetic (BF) phase, where both the bulk and the surface are magnetically ordered. For $T > T_c^{3D}$, the bulk is paramagnetic for all values of J_S/J_B, and the same happens with the surface if $\Delta \leq \Delta_c$ (with $\Delta_c \approx 0.5$-0.6). But if $\Delta > \Delta_c$, an interesting intermediate surface ferromagnetic (SF) region appears where surface magnetic order exists, even if bulk order is absent. The SF phase emerges for T in the interval $(T_c^{3D}, T_c^S(\Delta))$ where, because the presence of the bulk enhances the correlations between surface spins, $T_c^S(\Delta) \geq T_c^{2D} \equiv n^{2D}J_S/k_B$ (with $n^{2D} = 2.269...$); note that $T_c^S(\Delta_c) = T_c^{3D}$, and that necessarily $\Delta_c < n^{3D}/n^{2D}-1$ (this general inequality implies, for the present model, $\Delta_c < 0.988$). For $T > T_c^S(\Delta)$, the entire system is in the paramagnetic (P) phase.

Let us say a few words on the magnetization (M) profile: if $\Delta \ll \Delta_c$ ($\Delta \gg \Delta_c$), M monotonously increases (decreases) from $M_S(T)$ (surface magnetization) to $M_B(T)$ (bulk magnetization) while going from the surface to deep in the bulk, and, if $\Delta \approx \Delta_c$, M(T) presents a flat profile (M almost independent of the depth). In all cases, the bulk asymptotic value $M_B(T)$ is exponentially approached from the surface. It is essentially this fact which explains why the criticality associated with the SF phase should be the 2D one (i.e., the system behaves as being infinite in two dimensions and finite in the third one).

To illustrate the four different universality classes associated with the present system, recall the thermal critical behavior of M: (i) for all values of Δ, $M_B \propto (T_c^{3D} - T)^{\beta(3D)}$; (ii) for $\Delta < \Delta_c$, $M_S \propto (T_c^{3D} - T)^{\beta_1}$; (iii) for $\Delta = \Delta_c$ (SB multicritical point), $M_S \propto (T_c^{3D} - T)^{\beta(SB)}$; (iv) for $\Delta > \Delta_c$, $M_S \propto (T_c^S(\Delta) - T)^{\beta(2D)}$. We verify consequently the existence of four different critical exponents β. Finally, in the $(\Delta - \Delta_c) \to +0$ limit, we expect $T_c^S(\Delta)/T_c^{3D} - 1 \sim A\,(\Delta/\Delta_c - 1)^{1/\phi}$ which defines the crossover critical exponent ϕ and critical factor A.

Let us close this section by adding that we expect a fifth nontrivial singularity to be present in this problem for $\Delta > \Delta_c$, namely a soft singularity in $M_S(T)$ at $T = T_c^{3D}$ (when $M_B(T)$ vanishes): we are currently examining this point.

3. RG Approach: Potts Model.

The real-space RG approach we shall use in the present paper is illustrated here on the q-state Potts ferromagnet (q = 1 and 2, respectively, recover bond percolation and the spin 1/2 Ising model). The Hamiltonian is given by

$$\mathcal{H} = -q \sum_{\langle i,j \rangle} J_{ij}\, \delta_{\sigma_i,\sigma_j} \qquad (\sigma_i = 1,2,\ldots,q) \quad , \tag{2}$$

with $J_{ij} = J_S > 0$ if both i and j sites belong to the free surface, and $J_{ij} = J_B > 0$ otherwise; we introduce $K_{ij} \equiv J_{ij}/k_BT$. The RG recursive relations are obtained by imposing

$$\exp(-\mathcal{H}'_{12} / k_BT) = \operatorname*{Tr}_{3,4,\ldots N} \exp(-\mathcal{H}_{12\ldots N} / k_BT) \quad , \tag{3}$$

where $\mathcal{H}'_{12}$ and $\mathcal{H}_{12\ldots N}$, respectively, denote the Hamiltonians associated with a (renormalized) single bond (two sites) and a relatively large two-terminal cluster G (N sites). By imposing (3) we preserve all the equilibrium thermodynamical quantities, as it implies the preservation of the partition function. Naturally we have to impose (3) twice: one to obtain $K_B' = f(K_B)$ (by using the bulk cluster G_B), and one to obtain $K_S' = g(K_S,K_B)$ (by using the surface cluster G_S). The choice of (G_B,G_S) determines the particular RG approximation. We have used two diffeent choices: (i) very simple clusters of the Migdal-Kadanoff type [32] (qualitatively, but not always quantitatively, reliable as long as second-order phase transitions are concerned, i.e., roughly $q \le 3$), and (ii) recently introduced [33,41] clusters of a rather sophisticated shape (both qualitatively and quantitatively reliable for $q \le 3$). We shall denote the first choice by $RG^{(1)}$ (it uses clusters noted $G_B{}^{(1)}$ and $G_S{}^{(1)}$: see Figure 2 of [32]), and the second one by $RG^{(2)}$ (it uses clusters $G_B{}^{(2)}$ and $G_S{}^{(2)}$: see Figures 2 and 3 of [33]).

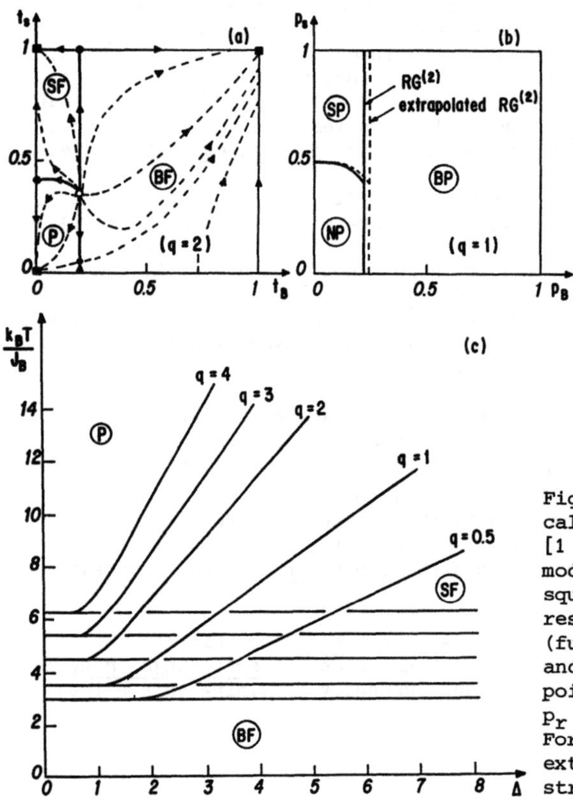

Figure 2. $RG^{(2)}$ phase diagram for typical values of q ($t_r \equiv [1 - e^{-qK_r}]/[1 + (q-1)e^{-qK_r}]$, $r = B,S$). (a) Ising model: the RG flow is indicated; full square, full circle, and open circle, respectively, denote the trivial (fully stable), critical (semi-stable), and multicritical (fully unstable) fixed points. (b) Bond percolation: $p_r \equiv 1 - e^{-K_r}$ ($r = B,S$) according to the Fortuin and Kasteleyn theorem; the extrapolation procedure ("horizontal" stretching providing, by construction, the best values available in the literature for t_c^{3D}) is indicated as well; note that the semi-infinite bulk makes *surface* percolation possible *below* the 2D threshold $p_S = 0.5$. (c) Standard k_BT/J_B vs. Δ representation.

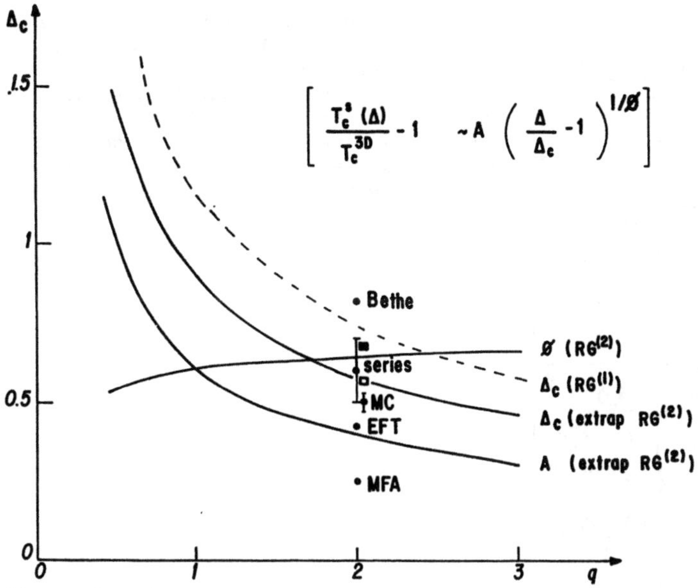

Figure 3. The q-evolution of Δ_c, A, and ϕ (our best values [33]). For q = 2, compare the present Δ_c (0.569) with other available results: mean-field approximation [9,10] (MFA; 0.25), effective field theory [15] (EFT; 0.4232), Bethe approximation [16] (0.816), $RG^{(1)}$ [32] (0.736), series [17] (0.6 ± 0.1), Monte Carlo [19] (MC; 0.50 ± 0.03). For q = 2, compare the present ϕ (0.641) with other available results: ε expansion [24] (full square; 0.68) and MC [19] (open square; 0.56 ± 0.04).

The mathematical operations involved in (3) can in principle be performed through the traditional (though tedious!) inspection of all microscopic spin configurations. But we have used instead the Break-collapse method (BCM) [42-44], which very conveniently solves (3) through simple topological operations. In fact, clusters such as those involved in $RG^{(2)}$ could hardly be solved, for arbitrary real value of q, were it not the BCM, as they yield polynomials of several thousands of terms. The knowledge of $f(K_B)$ and $g(K_S,K_B)$ closes the procedure as: (i) the RG flow in the (K_B,K_S) space determines the phase diagram as well as the universality classes, and (ii) the Jacobian $\partial(K_B',K_S')/\partial(K_B,K_S)$ at the relevant fixed points determines the values of the correlation length and crossover critical exponents (ν's and ϕ, respectively). The main results are indicated in Figures 2 and 3. The calculation of the values of the various critical exponents β appearing in the problem involves equations of state, and therefore a further step in complexity: we are curently working on that problem.

4. Mixed Ferro- and Antiferromagnetic Interactions.

Within the $RG^{(1)}$ framework we have studied [38] the influence, on criticality, of both signs for the q-state Potts coupling constants J_B and J_S. The situation described in Section 3 is herein recovered as the $J_B > 0$ and $J_S > 0$ particular case. The q-evolution of the phase diagram (in both t_B vs. t_S and T vs. J_S/J_B representations; see caption of Figure 2 for definitions of t_B and t_S) and associated universality classes is depicted in Figure 4. The well known ferro ↔ antiferro isomorphism in the simple cubic spin 1/2 Ising model is recovered for q = 2 [Figures 4(c) and (d)]. The phase diagram presents, besides the paramagnetic phase (P), various magnetic orders, namely, the bulk ferromagnetic (BF; simultaneous surface ferromagnetic order), the bulk antiferromagnetic (BAF; simultaneous surface antiferromagnetic order), the surface ferromagnetic (SF, the bulk is disordered), the surface antiferromagnetic (SAF; the bulk is disordered),

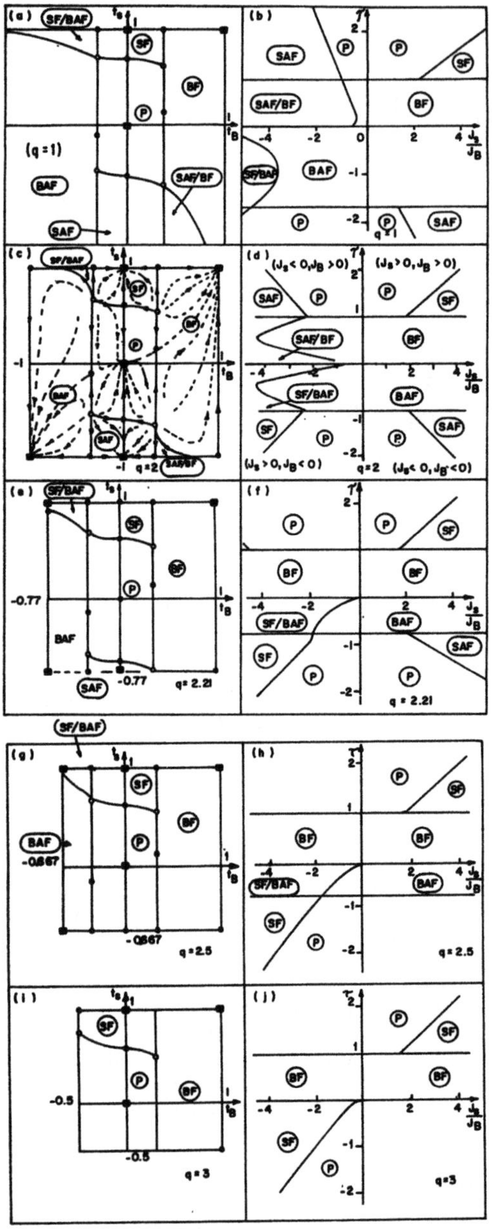

Figure 4. The q-evolution of the phase diagram; $\tau \equiv \mathrm{sign}(J_B)T/T_C$, where T_C is the bulk Curie temperature. Full squares, full circles and open circles, respectively, denote trivial, critical, and multicritical fixed points. The dashed lines are indicative. The $q=2$ case has been indicated with more details as a prototype.

the simultaneous surface ferromagnetic and bulk antiferromagnetic (SF/BAF), and the simultaneous surface antiferromagnetic and bulk ferromagnetic (SAF/BF) ones. The various critical and multicritical fixed points determine the corresponding universality classes. In particular, the RG flow on the SF/BAF-BAF and SAF/BF-BF critical lines suggests first-order phase transitions. Further inspection is needed to clearly establish this possibility.

Note that while q increases, the SAB/BF phase disappears, then the SAF phase, and finally the SF/BAF and the BAF phases disappear as well. The disappearance of the SAF phase is already rigorously established [45] for $t_B = 0$: note, however,

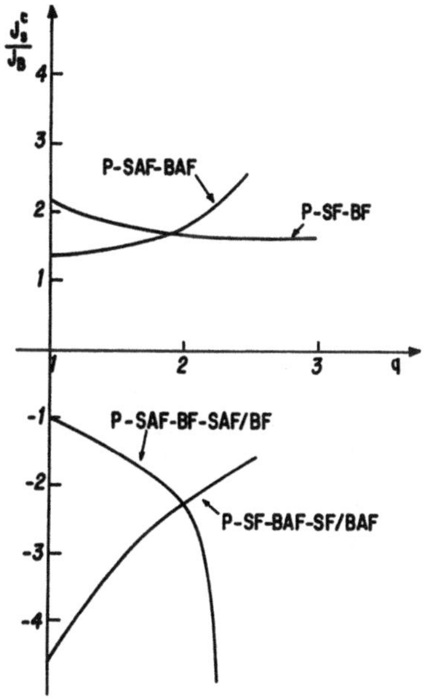

Figure 5. The q-evolution of $J_S{}^c/J_B$ (for $|J_S/J_B|$ above $|J_S{}^c/J_B|$ surface magnetic ordering can occur even in the absence of bulk ordering). For q = 2, $J_S{}^c/J_B$ = 1.736 (compared to the series result [17] 1.60, and the Monte Carlo one [19] 1.50) and -2.28 (compared to the series result [17] -1.9). For the simple-cubic lattice, the present RG cannot be retained much above q = 3 as the bulk transitions will become of the first order.

that the present RG approximation provides the disappearance at a value of q lower than it should (q = 2.25 instead of the exact [45] value 3). The q-evolution of the four multicritical points (J_B > 0 or < 0, J_S > 0 or < 0) is depicted in Figure 5.

Let us finally address the most interesting phenomenon exhibited by this analysis, namely the existence, for q ≈ 2, of re-entrances in the phase diagram. This is to say, for negative J_S/J_B and within the appropriate range (which depends on q), we have, while increasing T, that the surface magnetic order vanishes at a relatively low temperature, and reappears again below but close to the 3D critical temperature, and finally vanishes again above but close to the already mentioned 3D point. In other words, the bulk order acts, on the surface one, somehow similarly to a uniform external magnetic field acting on a (highly anisotropic) antiferromagnet. According to the above facts, the surface order parameter should present a maximum in the neighborhood of (possibly above) the 3D critical temperature: this is precisely what happens in Gd [8]! Although we are not aware of any concrete arguments relating Gd to the q ≈ 2 Potts model, both experimental and theoretical facts are striking enough to suggest that in Gd competitive signs might exist between surface and bulk coupling constants.

5. Interface Model.

The free surface case (bulk-surface-vacuum, with coupling constants J_B,J_S and 0, respectively) we have been discussing up to now can be generalized into the interface case (bulk 1 - surface - bulk 2, with coupling constants J_1, J_S, and J_2, respectively). Our calculations have been performed within the $RG^{(2)}$ approach, for the (1,0,0) interface between two semi-infinite simple cubic lattices assuming ferromagnetic Potts interactions. The RG flow diagram is illustrated, for q = 2, in Figure 6; its J_2/J_1 = 1/2 section is represented, in the (J_S/J_1,T) space, in Figure 7. We now have multicritical lines (which include the previous multicritical points) which join in a high-order multicritical point (noted equal bulk point), in the neighborhood of which a new type of crossover occurs. The parameter $\Delta_c \equiv J_S/J_1 - 1$ (above which surface order is possible in the absence of bulk order) monotonously decreases when q and/or J_2/J_1 increase (see Figure 8).

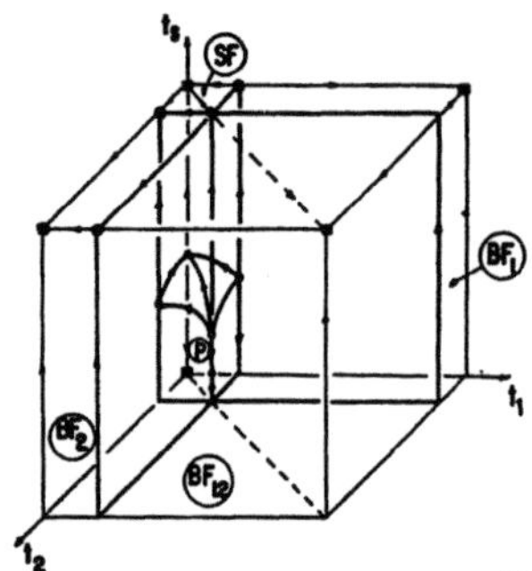

Figure 6. The q = 2 $RG^{(2)}$ flux diagram; full squares, full circles and open circles denote, respectively, fully stable (trivial), semistable (critical and multicritical), and fully unstable (high-order multicritical) fixed points. The five possible phases are indicated: the two single-bulk ferromagnetic (BF_1 and BF_2), the double-bulk ferromagnetic (BF_{12}), the surface ferromagnetic (SF), and the paramagnetic (P) phases.

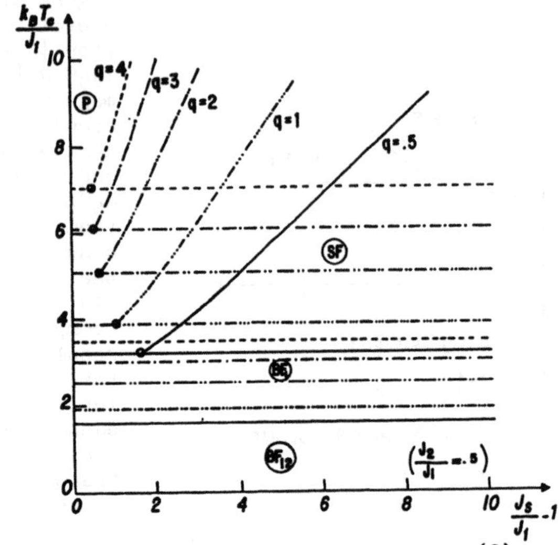

Figure 7. The q-evolution in $RG^{(2)}$ the phase diagram for $J_2 / J_1 = 0.5$.

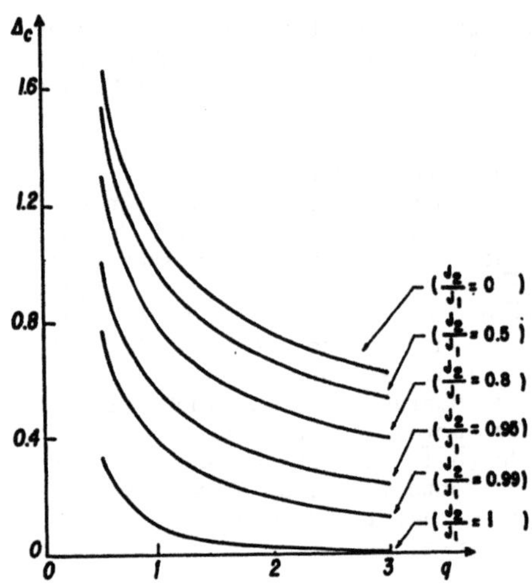

Figure 8. The q-evolution in $RG^{(2)}$ of Δ_c for typical ratios J_2/J_1 ($J_2/J_1 = 0$ and 1, respectively, correspond to the *free surface* and *equal bulks* cases).

6. Symmetry of the Interaction.

It is well known that the symmetry of the interaction is a very relevant ingredient of critical phenomena (e.g., it characterizes the universality class at a given dimensionality D). This is particularly true for D = 2, where the Mermin-Wagner theorem forbids the existence of long-range order at any finite temperature if the (short-range) interaction is invariant under a continuous group of symmetry (e.g., isotropic Heisenberg and XY models, in opposition to the Ising model, which is related to a discrete group of symmetry). Let us also recall that the theorem says nothing about the existence of a phase transition: indeed, the D = 2 XY model presents the well known Kosterlitz-Thouless phase transition, detectable through a divergence of the susceptibility. The Heisenberg model has a continuous group of symmetry which is larger than that of the XY model, and consistently presents, at finite temperatures, no phase transition at all.

We go now back to the free surface (or even the interface) problem we are dealing with. Within the temperature interval where the SF phase exists, the magnetization profile is expected, as already said, to vanish exponentially while penetrating into the paramagnetic bulk(s). Its criticality is therefore expected to be that of the 2D system, and naturally all above considerations should hold. It is on these grounds that it becomes quite interesting the systematic analysis of the following spin 1/2 anisotropic Heisenberg Hamiltonian:

$$\mathcal{H} = -\sum_{\langle i,j\rangle} J_{ij} \left[(1 - \eta_{ij}) (\sigma_i^x \sigma_j^x + \sigma_i^y \sigma_j^y) + \sigma_i^z \sigma_j^z \right] \quad , \qquad (4)$$

where (J_{ij}, η_{ij}) equals (J_S, η_S) if both i and j sites belong to the free surface, and equals (J_B, η_B) otherwise (note that $\eta_{ij} = 1, 0$ and $-\infty$, respectively, correspond to the Ising, isotropic Heisenberg, and isotropic XY models). The $J_B, J_S > 0$, $\eta_B = 1$, and $0 \leq \eta_S \leq 1$ models have been analyzed [34] within the RG$^{(1)}$ approach (the RG procedure which enables the treatment of quantum systems like the present one is described in Reference 46). The RG flow diagram and the η_S-dependence of Δ_c are respectively presented in Figures 9 and 10(a). The extension to the interface problem ($J_2 \neq 0$ and $J_1 \equiv J_B$) has been discussed [37] as well: the results are depicted in Figure 10. The parameter Δ_c monotonously increases when η_S and/or J_2/J_1 vary from 1 to 0. However, it remains finite even for $\eta_S = J_2/J_1 = 0$: this is not in contradiction with the Mermin-Wagner theorem (oe even with the fact that $T_c = 0$ for the D = 2 isotropic Heisenberg model) because $\eta_B = 1$, and consequently not all the interactions present continuous groups of symmetry. We are currently calculating [47] the influence of η_B, and will hopefully obtain $\lim_{\eta_B, \eta_S \to 0} \Delta_c = \infty$, consistently with the D = 2 critical peculiarities.

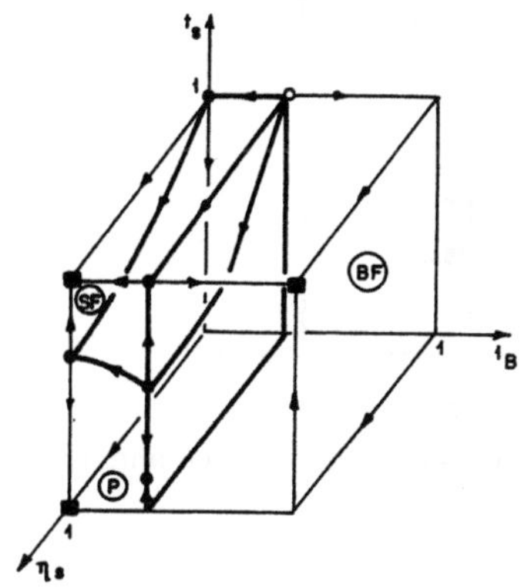

Figure 9. The flux diagram in RG$^{(1)}$ for the free-surface anisotropic Heisenberg on Ising bulk ferromagnet ($t_r \equiv \tanh K_r$, r = B,S); $\eta_S = 1$ corresponds to the Ising model.

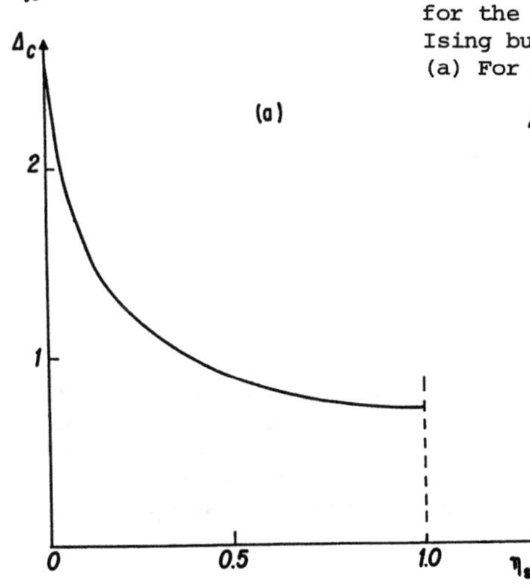

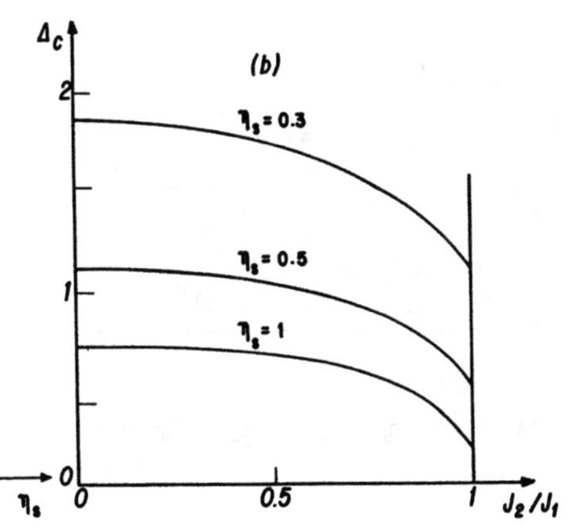

Figure 10. Dependence of Δ_c on η_S and J_2J_1 in RG$^{(1)}$ for the interface anisotropic Heisenberg between two Ising bulks; $\eta_S = 1$ corresponds to the Ising model. (a) For $J_2/J_1 = 0$; (b) for typical values of η_S.

7. Bond-Dilute Model.

Here we consider a different type of extension, namely, quenched bond-dilution. The Potts ferromagnetic coupling constants for the interface problem will be assumed to be random variables with the following probability laws:

$$P_r(J_{ij}) = (1 - p_r)\,\delta(J_{ij}) + p_r\,\delta(J_{ij} - J_r) \qquad (r = S,1,2) \quad . \tag{5}$$

The $p_S = p_1 = p_2 = 1$ particular case recovers the pure model we have been considering in Section 5. The phase diagram of this quite general system involves hypersurfaces in a six-dimensional parameter space, e.g., $(k_BT/J_1,\ J_S/J_1,\ J_2/J_1,\ p_S,\ p_1,\ p_2)$, the study of which is currently in progress. A few results within the $RG^{(1)}$ framework (bulk dilution in the q = 2 free surface problem [36], interface dilution for arbitrary q [39], simultaneous interface and bulk dilution for arbitrary q [40]), as well as within the $RG^{(2)}$ one (free surface dilution for arbitrary q [48]), are already available. Some interesting effects are depicted in Figures 11 and 12. In particular, we see in Figure 11 that bulk dilution might be an excellent experimental manner for making appear, at a given value of J_S/J_B (which is fixed for a given substance), surface magnetism.

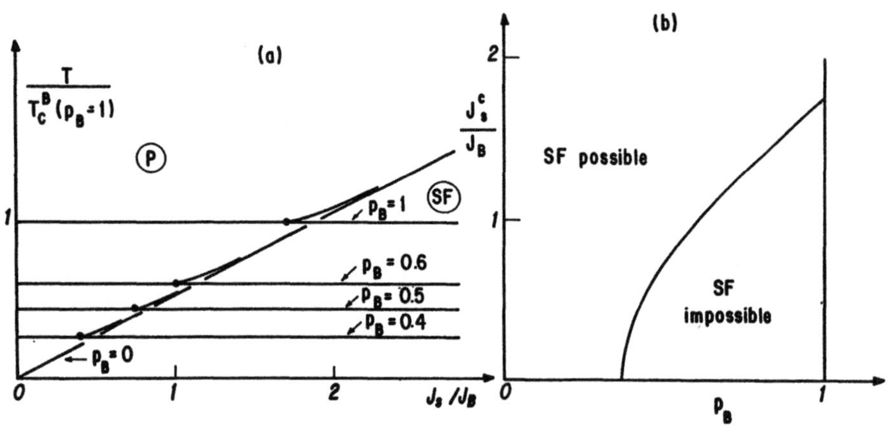

Figure 11. Free surface Ising $RG^{(1)}$ p_B-evolution of the phase diagram (a) and the localization $J_S{}^c/J_B$ of the multicritical point (b).

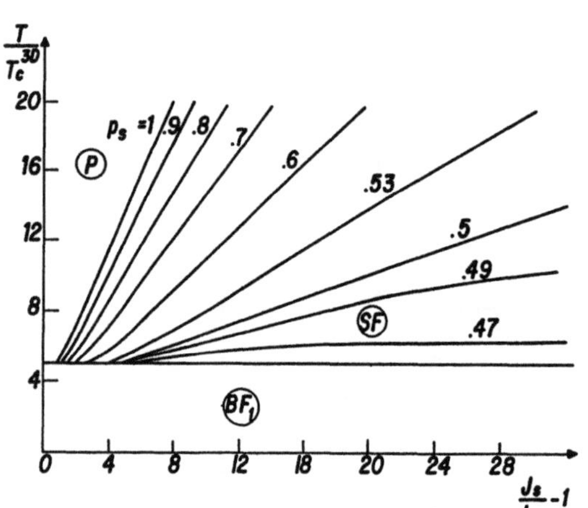

Figure 12. Free surface Ising $RG^{(2)}$ p_S- evolution of the phase diagram. Note the existence of the SF phase *below* the 2D percolation threshold $p_S = 1/2$; it disappears at $p_S = 0.415 \pm 0.003$.

8. Conclusion.

The influence of several factors on surface magnetism has been tutorially reviewed. Among those which present interesting effects and which might have strong relevance for experimental work, let us select the following ones: (i) ferro-antiferro competition between surface and bulk coupling constants; (ii) symmetry of the surface and bulk interactions; (iii) bulk dilution. Experimental evidence on these and other effects discussed here would be extremely welcome and enlightening.

References.

1. D. T. Pierce and F. Meier, Phys. Rev. B 13, 5484 (1976).
2. R. J. Celotta, D. T. Pierce, G. C. Wang, S. D. Bader, and G. P. Fercher, Phys. Rev. Lett. 43, 728 (1979).
3. C. Rau and S. Eichner, in Nuclear Methods in Materials Research, ed. by K. Bedge, H. Bauman, H. Jex, and F. Rauch (Vieweg, Braunschweig, 1980), p. 354.
4. S. F. Alvarado, M. Campagna, and H. Hopster, Phys. Rev. Lett. 48, 51 (1982).
5. S. F. Alvarado, M. Campagna, F. Ciccacci, and H. Hopster, J. Appl. Phys. 53, 7920 (1982).
6. C. Rau, J. Magn. Magn. Mater. 31-34, 874 (1983).
7. L. E. Klebanoff, S. W. Robey, G. Liv, and D. A. Shirley, Phys. Rev. B 30, 1048 (1984).
8. D. Weller, S. F. Alvarado, W. Gudat, K. Schröder, and M. Campagna, Phys. Rev. Lett. 54, 1555 (1985).
9. D. L. Mills, Phys. Rev. B 3, 3887 (1971).
10. D. L. Mills, Phys. Rev. B 8, 4424 (1973).
11. F. Aguilera-Granja and J. L. Morán-López, Phys. Rev. B 31, 7146 (1985).
12. E. F. Sarmento, R. Honmura, I. P. Fittipaldi, and T. Kaneyoshi, Proceedings of the 2nd Latin-American Symposium on Surface Physics (Mexico), ed. by P. Halevi (Puebla, Mexico, 1982).
13. I. Tamura, E. F. Sarmento, I. P. Fittipaldi, and T. Kaneyoshi, Phys. Stat. Sol. (b) 118, 409 (1983).
14. T. Kaneyoshi, I. Tamura, and E. F. Sarmento, Phys. Rev. B 28, 6491 (1983).
15. E. F. Sarmento, I. Tamura, and T. Kaneyoshi, Z. Phys. B 54, 241 (1984).
16. F. Aguilera-Granja, J. L. Morán-López, and J. Urías, Phys. Rev. B 28, 3909 (1983).
17. K. Binder and P. C. Hohenberg, Phys. Rev. B 9, 2194 (1974); P. C. Hohenberg and K. Binder, AIP Conf. Proc. 24, 300 (1975).
18. S. Selzer and N. Majlis, Phys. Rev. B 27, 544 (1983).
19. K. Binder and D. P. Landau, Phys. Rev. Lett. 52, 318 (1984).
20. T. W. Burkhardt and E. Eisenriegler, Phys. Rev. B 16, 3213 (1977).
21. N. M. Svrakic and M. Wortis, Phys. Rev. B 15, 396 (1977).
22. T. W. Burkhardt and E. Eisenriegler, Phys. Rev. B 17, 318 (1978).
23. N. M. Svrakic, R. Pandit, and M. Wortis, Phys. Rev. B 22, 1286 (1980).
24. H. W. Diehl and S. Dietrich, Phys. Lett. 80A, 408 (1980).
25. R. Lipowsky and H. Wagner, Z. Phys. B 42, 355 (1981).
26. H. W. Diehl and S. Dietrich, Z. Phys. B 42, 65 (1981).
27. R. Lipowsky, J. Phys. A 15, L195 (1982).
28. R. Lipowsky, Z. Phys. B 45, 229 (1982).
29. M. Wortis and N. M. Svrakic, IEEE Trans. Mag. Mag. 18, 721 (1982).
30. P. M. Lam and Z. Q. Zhang, Z. Phys. B 52, 315 (1983); errata: Z. Phys. B 55, 371 (1984).
31. K. Binder, "Critical behavior at surfaces," in Phase Transitions and Critical Phenomena, ed. by C. Domb and J. L. Lebowitz, Vol. 8 (Academic Press, 1983).
32. C. Tsallis and E. F. Sarmento, J. Phys. C 18, 2777 (1985).
33. U. M. S. Costa, C. Tsallis, and E. F. Sarmento, J. Phys. C 18, 5749 (1985).
34. U. M. S. Costa, A. M. Mariz, and C. Tsallis, J. Physique Lett. 46, L851 (1985) (erratum: at the ordinate origin of Figure 4 read 0.75 instead of 0).
35. L. R. da Silva, U. M. S. Costa, and C. Tsallis, J. Phys. C 18, 6199 (1985).
36. C. Tsallis, E. F. Sarmento, and E. L. Albuquerque, J. Magn. Magn. Mat. 54-57, 667 (1986).

37. R. J. V. dos Santos, E. F. Sarmento, C. Tsallis, and U. M. S. Costa, Phys. Rev. B 33, 1741 (1986); an erratum is being published concerning the values of Δ_c for $\eta \ll 1$.
38. E. F. Sarmento and C. Tsallis, J. Phys. (Paris) 47, 1115 (1986).
39. S. B. Cavalcanti and C. Tsallis, private communication and J. Phys. C, to appear in 1986.
40. E. F. Sarmento, E. L. Albuquerque, and C. Tsallis, private communication and to be published.
41. L. R. da Silva, C. Tsallis, and G. Schwachheim, J. Phys. A 17, 3289 (1984).
42. C. Tsallis and S. V. F. Levy, Phys. Rev. Lett. 47, 950 (1981).
43. J. W. Essam and C. Tsallis, J. Phys. A 19, 411 (1986).
44. A. C. N. de Magalhães and J. W. Essam, J. Phys. A 19, 1655 (1986).
45. R. J. Baxter, Proc. R. Soc. London Ser. A 383, 43 (1982).
46. A. O. Caride, C. Tsallis, and S. I. Zanette, Phys. Rev. Lett. 51, 145 (1983); erratum: Phys. Rev. Lett. 51, 616 (1983); A. M. Mariz, C. Tsallis, and A. O. Caride, J. Phys. C 18, 4189 (1985).
47. A. M. Mariz, U. M. S. Costa, and C. Tsallis, private communication and to be published.
48. L. R. da Silva, C. Tsallis, and E. F. Sarmento, private communication and to be published.

Monte Carlo Studies of Two-Dimensional Ising Models with Competing Interactions

D.P. Landau

Department of Physics and Astronomy, University of Georgia,
Athens, CA 30601, USA

1. Introduction.

Beginning with Onsager's solution of the nearest neighbor Ising square lattice [1], the Ising model has served as a testing ground for a host of different theoretical approaches as well as a guide to experimentalists studying phase transitions in pseudo-two-dimensional magnetic systems. When we allow for the possibility of competing interactions or the application of a magnetic field, the range of phase transitions and critical behavior becomes exceedingly rich. In most cases, however, these models are not amenable to exact solution. Monte Carlo methods have proven to be quite valuable in the study of a wide range of two-dimensional models, and we shall review some of this work in this paper. Possible physical manifestations of these models include magnetic monolayers, layered magnetic compounds, as well as magnetic graphite intercalates and magnetic surfaces with enhanced exchange which are discussed elsewhere in this volume.

2. Background.

A general form for the Ising Hamiltonian is given by

$$\mathcal{H} = \sum_{(ij)} J_{ij}\,\sigma_i\,\sigma_j + \sum_{(ijk)} P_{ijk}\,\sigma_i\,\sigma_j\,\sigma_k + H\sum_i \sigma_i \quad , \tag{1}$$

where $\sigma = \pm 1$. The first summation is over all pairs of the lattice with a given pair interaction, and the second sum is over all interacting triangles. (This latter term is unimportant for magnetic models but can play an important role in equivalent lattice-gas models for adsorbed monolayers. We shall discuss this equivalence shortly.) The effect of including a magnetic field H is also important and is given by the third term in (1). This "generalized" Ising model has been treated using many different approaches with varying degrees of success. Mean-field theory has been used widely but is known, in some cases, to yield qualitatively incorrect results, and in other cases the role of fluctuations appears to change the quantitative nature of the critical behavior.

An important extension of the simple two-dimensional Ising model is the three-dimensional Ising model with free surfaces. Here we are primarily interested in phase transitions and critical behavior occurring in the surface layer and the manner in which they are affected by the presence of the underlying bulk. An appropriate Hamiltonian for this model is then given by

$$\mathcal{H} = J \sum_{\substack{(ij) \\ \text{in bulk}}} \sigma_i\,\sigma_j + J_s \sum_{\substack{(ij) \\ \text{at surface}}} \sigma_i\,\sigma_j \quad , \tag{2}$$

where the exchange J_s in the surface layer may be different from that in the bulk J.

In recent years increasing attention has been paid to the study of phase transitions in adsorbed monolayers on crystalline substrates. In fact an exceedingly good description is provided by a lattice-gas model which can easily be shown to be completely equivalent to an Ising model [2]. If we consider the adatoms to be localized on sites of a lattice defined by the underlying substrate, we can introduce site occupation variables c_i where $c_i = 1$ if a site is occupied and $c_i = 0$ if the site is empty. Each adsorbed atom is bound to the substrate with energy ε, and the entire system is in equilibrium with the surrounding vapor described by chemical potential μ. The appropriate grand canonical potential is then

$$- \mu N_a = \sum_{(ij)} \phi_{ij} c_i c_j + \sum_{(ijk)} \phi_{ijk} c_i c_j c_k + (\varepsilon + \mu) \sum_i c_i \quad , \tag{3}$$

where the ϕ's are interaction energies between adatoms. N_a is the number of adatoms and μ is the chemical potential. Using the simple transformation $c_i = (1-\sigma_i)/2$ we can easily show that this is completely equivalent to a simple Ising model in a magnetic field H where

$$\begin{aligned} J_{ij} &= (1/4)\, \phi_{ij} + (1/8) \sum_{k(\neq ij)} \phi_{ijk} \quad , \\ P_{ijk} &= (1/8)\, \phi_{ijk} \quad , \\ - H &= -(1/2)\,(\varepsilon + \mu) + (1/4) \sum_{j(\neq i)} \phi_{ij} + (1/8) \sum_{j \neq k(\neq i)} \phi_{ijk} \quad . \end{aligned} \tag{4}$$

Note that this equivalent Ising model is treated in a canonical ensemble and that a lattice gas model in the canonical ensemble is the same as an Ising model at constant magnetization.

3. Monte Carlo Method.

Since several complete descriptions of the application of Monte Carlo methods to the study of statistical mechanics systems are in the literature [3], we shall only briefly review them here. The equilibrium value of any parameter A of a system can in principle be calculated exactly from:

$$\langle A \rangle = Z^{-1} \sum_{\mu} A_{\mu} \exp(-E_{\mu}/k_B T) \quad , \tag{5}$$

where A_μ is the value of the parameter in the μth configuration, E_μ is the energy of that state, and Z is the partition function.

Unfortunately, the number of terms in the summations becomes astronomical for even systems of modest size. Instead we can sample the states of the system and approximate the desired result by an average over the states actually sampled. This procedure in its simplest form is rather imprecise, and instead it is far better to sample states with a probability which is proportional to their contribution to the partition function. In practice this is done by considering each site in turn and overturning spins with probability

$$p = \exp(-\Delta E/k_B T) \quad , \tag{6}$$

where ΔE is the cost of energy due to flipping.

In this manner a Markov chain of states is produced and an estimate for $\langle A \rangle$ is produced by a simple arithmetic average over the states generated. Since successive states are correlated, two different time-dependent effects may occur. First of all the system may require a number of MCS (Monte Carlo steps per site, i.e. sweeps through the lattice) to reach equilibrium, and thus the first configurations generated must be discarded. In addition, near a phase transition, critical slowing down occurs and produces long relaxation times. One last important "technical" consideration is the finite lattice size. The effect of finite lattice size is now well understood, and the infinite lattice critical behavior may be extracted using finite size scaling theory.

4. Phase Diagrams.

The most straightforward Ising models are those for which $J_{ij} < 0$, i.e. the interactions are all ferromagnetic. If this is the case, the ordered state is ferromagnetic, but the transition temperature depends on the details of the interactions. Whenever some of the interactions are antiferromagnetic, a much richer set of phase diagrams are possible.

As we show in Figure 1(a), the ordered state may change as a function of the relative size of nnn- and 3nn- coupling [4]. It is particularly interesting to note that for the specific value of $R = J_{nnn}/J_{nn}$ used to plot Figure 1(b) none of the ordered states exhibit phase transitions with two-dimensional Ising critical exponents. The behavior of the square lattice has also been determined for $R > 0$ and $R' = J_{3nn}/J_{nn} = 0$. For $R < 0$ a tricritical point has been observed on the field-induced phase boundary, and for $R > 0$ phase boundaries have been found which have non-universal behavior. In the presence of sufficiently strong magnetic fields the groundstate for $R > 0$ may appear to become 2×2; however, in the absence of a 3nn coupling it is possible to shift randomly either rows or columns (but not both) with no cost in energy thus destroying long-range order.

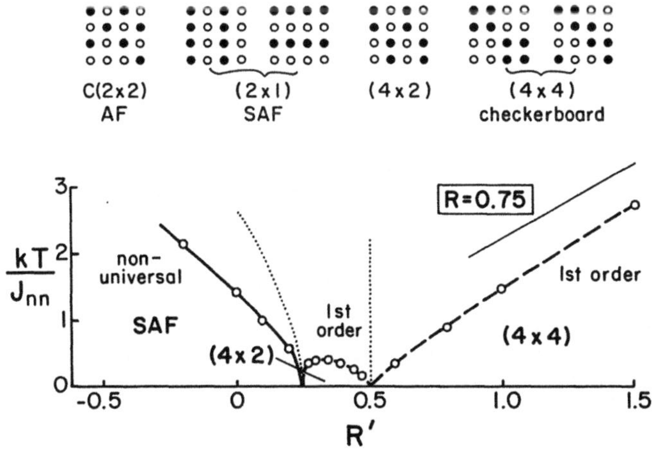

Figure 1. (upper part) Possible ordered states for the Ising square lattice: Filled circles are spin-down and open circles are spin-up.
(lower part) Phase diagram as a function of 3nn-coupling ($R' = J_{3nn}/J_{nn}$).

The phase diagrams for the triangular lattice are also quite interesting. For $R = 0$ and $R' = 0$ there is no phase transition at any finite temperature in zero field. With the application of a magnetic field a ferrimagnetic $(\sqrt{3} \times \sqrt{3})^*$ ordered state becomes stable and it is separated from the disordered state by a 3-state Potts-like phase boundary [5]. The phase diagram is symmetric about $H = 0$ and a $(\sqrt{3} \times \sqrt{3})^*$ appears for $H < 0$. (For $P \neq 0$ the phase diagram becomes asymmetric and the $(\sqrt{3} \times \sqrt{3})^*$ phase eventually disappears. As P is varied, multicritical or critical endpoints appear [6].) With the addition of ferromagnetic nnn-coupling ($R < 0$) tricritical points appear on the phase boundary for non-zero field and the zero field behavior also changes dramatically (see Figure 2) [7]. The specific heat, shown in Figure 3, shows two finite maxima. The high temperature susceptibility shows an exponential divergence at critical temperature T_1, which is just below the upper specific heat peak, and a non-zero order parameter appears below T_2, which is on the upper side of the lower specific heat peak. A thorough analysis of the bulk properties, the correlation length, and the observation of vortex-like excitations produce strong evidence that there is an xy-like state separating a low-temperature ordered phase from a high-temperature disordered

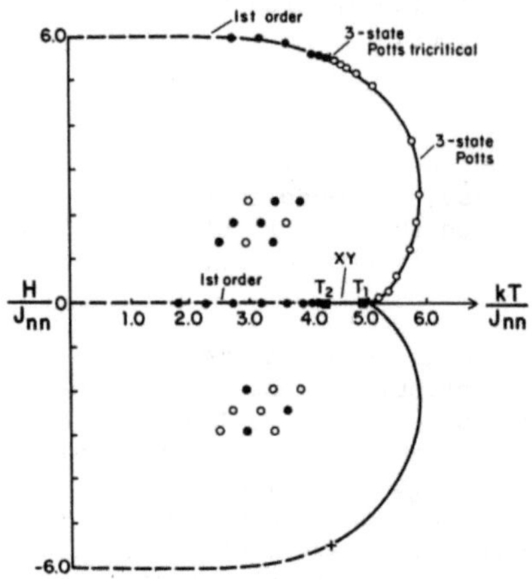

Figure 2. Field-temperature phase diagram for the triangular Ising model with $J_{nn} > 0$ and $R = -1$.

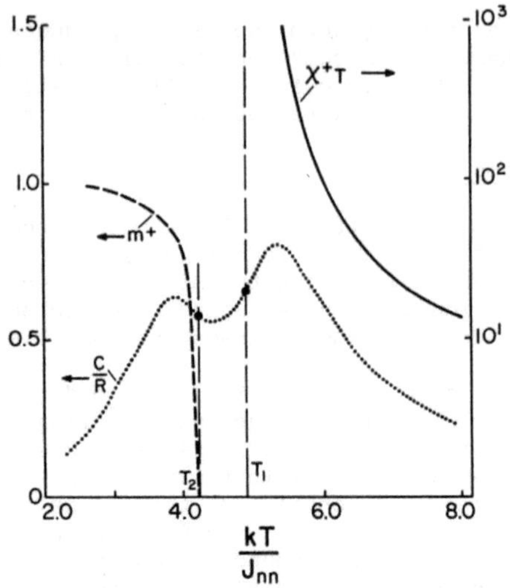

Figure 3. Bulk properties for the triangular Ising model with $J_{nn} > 0$ and $R = -1$.

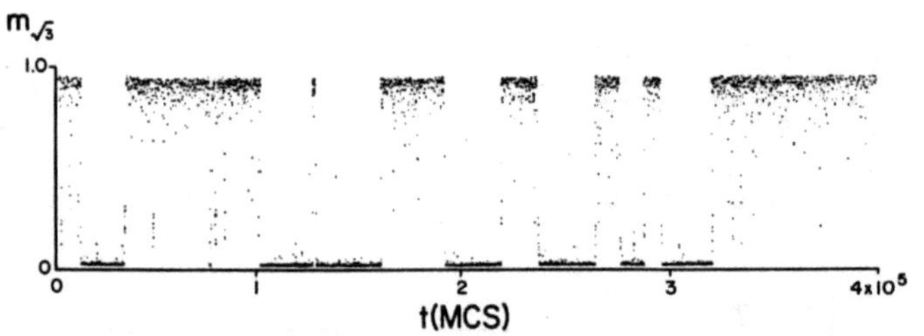

Figure 4. Time dependence of the order parameter for the $(\sqrt{3} \times \sqrt{3})$ groundstate for the triangular Ising model with $J_{nn} > 0$, $R = -1$, and $R' = -1$. The temperature is $T \approx 0.98\ T_c$.

state. This result is in agreement with the theoretical prediction of JOSÉ et al. [8] for the behavior of the xy-model with sixth order anisotropy. With the addition of a ferromagnetic 3nn coupling of the same strength, the transition becomes first-order [9]. As shown in Figure 4 the time dependence of the $(\sqrt{3} \times \sqrt{3})$ order parameter shows long-lived ordered states with jumps to a state with no $(\sqrt{3} \times \sqrt{3})$ long range order. Observation of other order parameters shows, however, that the state to which the system jumps is not disordered but has (2×2), (2×1), or (1×1) order. Thus the ordered state appears to remain highly degenerate.

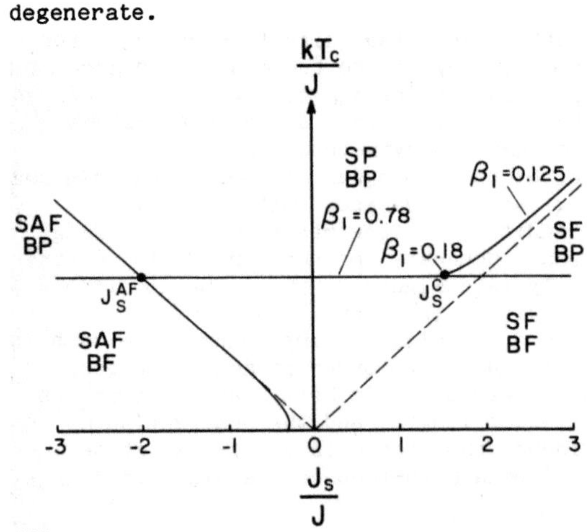

Figure 5. Phase diagram for the simple cubic Ising model with ferromagnetic bulk exchange J and surface exchange J_s. All phases are labeled by both the bulk state (e.g. BF is bulk, ferromagnetic) and the surface state (e.g. SAF is surface, antiferromagnetic).

The last example which we shall consider is that of a semi-infinite Ising model with interaction J everywhere but in the surface layer itself where the value J_s may be decidedly different [10]. In Figure 5 we show the resultant phase diagram where we label both bulk and surface order in all states. Along the phase boundary where the bulk and surface disorder simultaneously, the surface order disappears with an exponent β_1 which is not the same as the two-dimensional Ising value. For $J > J^c_s$ the surface stays ordered above the bulk transition temperature and shows ordinary Ising critical behavior. At the bulk-surface multicritical point J^c_s the exponents assume still different values. The antiferromagnetic surface-bulk multicritical point J^{AF}_s is a decoupled tetracritical point, and the surface and bulk essentially ignore each other.

5. Conclusion.

With the inclusion of competing interactions the Ising model becomes an interesting subject for further investigation. The Monte Carlo method has proven very well suited for these studies, and it would be particularly interesting to see some of the observed phenomena in real, physical systems.

Acknowledgments

Portions of this work were carried out in collaboration with K. Binder, K. K. Chin, and R. Subramaniam. This research was supported in part by U.S. National Science Foundation and by the North Atlantic Treaty Organization.

References

1. L. Onsager, Phys. Rev. 65, 117 (1944).
2. For a full discussion of the transcription see K. Binder and D. P. Landau, Surf. Sci. 108, 503 (1981).
3. See, e.g., Monte Carlo Methods in Statistical Physics, ed. K. Binder (Springer-Verlag, Berlin, Heidelberg, New York, 1979).
4. D. P. Landau and K. Binder, Phys. Rev. B 31, 5946 (1985).
5. B. D. Metcalf, Phys. Lett. 45A, 1 (1973); M. Schick, J. S. Walker, and M. Wortis, Phys. Rev. B 16, 2205 (1977).
6. K. K. Chin, Thesis, U. of Georgia, 1982 (unpublished), and K. K. Chin and D. P. Landau (to be published).
7. D. P. Landau, Phys. Rev. B 27, 5604 (1983).
8. J. V. José, L. P. Kadanoff, S. Kirkpatrick, and D. R. Nelson, Phys. Rev. B 16, 1217 (1977).
9. R. Subramaniam, Thesis, U. of Georgia, 1984 (unpublished); R. Subramaniam and D. P. Landau (to be published).
10. K. Binder and P. C. Hohenberg, Phys. Rev. B 9, 2194 (1976); K. Binder and D. P. Landau, Phys. Rev. Lett. 52, 318 (1984); Surf. Sci. 151, 409 (1985).

Critical and Multicritical Phenomena in Semi-Infinite Ising Systems

J.M. Sánchez[1] *and J.L. Morán-López*[2]

[1]Henry Krumb School of Mines, Columbia University, New York, NY 10027, USA
[2]Departamento de Física, Centro de Investigacion y de Estudios Avanzados del I.P.N., Apartado Postal 14-740, México, D.F., México

The magnetic behavior of semi-infinite solids and their thin films have been extensively studied using Monte Carlo [1,2], renormalization-group [3,4], and mean-field [5-7] approximations. These calculations have been limited to bulk systems with simple cubic structures and to the surface properties of the (100) plane. In particular, the recent Monte Carlo calculations of BINDER and LANDAU [2] have provided very reliable values of the critical exponents for the surface magnetization as well as for the value of surface exchange interaction J_c beyond which the surface critical temperature is higher than the bulk. Concerning critical exponents, those estimated for the simple cubic lattice are expected to be applicable to other crystal structures. However, the critical value of the surface exchange interaction would depend strongly on the particular crystal structure and on the surface of interest. At present J_c is unknown for experimentally more relevant structures, such as the fcc.

In this work we investigate a semi-infinite fcc lattice with a (100) surface using a cluster extension of the molecular-field approximation [8,9]. Although this mean-field theory strictly predicts classical critical behavior, it is well established that it nevertheless provides accurate values for the critical temperature in bulk systems [10]. Here we show that it can also be applied reliably to the computation of critical temperatures in semi-infinite systems. As an example, the surface critical behavior of the semi-infinite cubic lattice is revisited. It is seen that this lattice treated in the cube approximation can be described within the accuracy achieved by the best Monte Carlo results to date. It is also seen that the calculated magnetizations, outside a relatively small classical critical region, are consistent with the known critical exponents. This agreement is expected from the fact that small cluster approximations are able to reproduce exactly a large number of terms in the high and low temperature expansions of the zero-field magnetic susceptibility and magnetization, respectively. For example, all known low temperature expansion coefficients of the magnetization in the fcc lattice are reproduced exactly using a six-point cluster [11].

In order to determine the equilibrium properties of the system at finite temperature, we use a generalization of the self-consistent mean-field approximation which, as discussed elsewhere [12], leads to the same level of approximation as the cluster variation method.

The cluster method consists in determining a set of reduced density matrices ρ_α which describe the probability of observing all possible arrangements (or configurations) of A and B atoms on a set of lattice sites (or clusters) $\alpha = \{i,j,\ldots k\}$. The reduced density matrices ρ_α may be written as:

$$\rho_\alpha = Z_\alpha^{-1} \exp(-H_\alpha/kT) \quad , \tag{1}$$

where Z_α is a normalization constant and H_α is the renormalized energy of cluster α. The reduced density matrices for all subclusters β of a given cluster α are given by the following consistency relations:

$$\rho_\beta = \mathrm{Tr}^{(\alpha-\beta)}\rho_\alpha \quad , \tag{2}$$

where the trace operator $Tr^{(\alpha-\beta)}$ stands for the sum over all configurational variables σ_i in cluster α with the exception of those in β.

In order to describe functions that depend on the cluster configuration, such as the renormalized Hamiltonians H_α, it is convenient to introduce a set of characteristic functions ϕ_α which are orthogonal with respect to the inner product in configuration space [12]. For binary systems, the characteristic functions are given by the product of the occupation operators σ_i at each lattice site in the cluster α:

$$\phi_\alpha = \prod_{\substack{i \\ (\text{in } \alpha)}} \sigma_i \quad . \tag{3}$$

The characteristic functions defined by (3), together with the characteristic function of the empty cluster $\phi_o = 1$, obey the following:

$$\langle \phi_\alpha, \phi_\beta \rangle = 2^{-n} \, Tr(\phi_\alpha \phi_\beta) = \delta_{\alpha\beta} \quad , \tag{4}$$

where Tr operates on all configurational variables σ_i in α and β, n is the total number of lattice sites in the cluster which is the union of α and β, and $\delta_{\alpha\beta}$ equals 1 if α is equal to β and zero otherwise. (Note that geometrically identical clusters, such as two tetrahedra, but which are located in different regions in the lattice are considered different clusters.)

Using the characteristic functions ϕ_α as an orthogonal basis in configuration space, we may write the renormalized Hamiltonian for cluster α as:

$$H_\alpha = \sum_{\substack{\beta \\ (\text{all } \beta \text{ in } \alpha)}} h_\beta^{(\alpha)} \phi_\beta \quad , \tag{5}$$

where $h_\beta^{(\alpha)}$ is an effective renormalized multisite interaction for β clusters in α. From (1), (4), and (5), it follows that:

$$h_\beta^{(\alpha)} = \langle \phi_\beta, H_\alpha \rangle = -k_B T \, \langle \phi_\beta, \ln\rho_\alpha \rangle \quad . \tag{6}$$

The main objective of the self-consistent cluster method is to determine the renormalized Hamiltonians H_β (or alternatively the renormalized multisite interactions $h_\beta^{(\gamma)}$) for a selected set of maximum clusters γ. Once the renormalized Hamiltonians of the maximum clusters are known, the reduced density matrices for all subclusters [see (2)] as well as the free energy of the system can be easily calculated [12].

Within the self-consistent cluster method, approximate expressions for the renormalized Hamiltonians H_α are obtained as follows. First, a set of irreducible cluster energies $\tilde{H}_\beta$ are defined such that, for all clusters α, the renormalized Hamiltonians can be written as:

$$H_\alpha = \sum_{\substack{\beta \\ (\text{all } \beta \text{ in } \alpha)}} \tilde{H}_\beta \quad , \tag{7}$$

where the sum is carried out over all clusters β contained in α. The Möbius transformation (7) between the renormalized and irreducible Hamiltonians is unique. Second, an approximate closure condition is introduced by assuming that the irreducible cluster energies vanish identically for all clusters larger than the maximum cluster γ. This approximation allows us to write the Hamiltonian of the system as [12]:

$$H = \sum_\alpha{}' \, a_\alpha H_\alpha \quad , \tag{8}$$

where the prime indicates that the sum includes only those clusters in the lattice which are contained in the set of maximum clusters γ. By application of (7) to the whole crystal, in which case the renormalized Hamiltonian equals the total Hamiltonian H, and comparison with (8), the following set of relations for the coefficients a_α are obtained:

$$\sum_{\alpha}{}' a_\alpha = 1 \quad , \tag{9}$$

(all α which contain β)

where the equations are valid for all clusters β contained in the set of maximum clusters. The coefficients a_α depend only on the nature of the lattice and can be obtained recursively from (9).

For an infinite crystal, the coefficients given by (9) are independent of the location of ther cluster, and will be simply denoted by a_n, where the index n refers to the number of sites in the cluster. In the case of the simple-cubic lattice, and taking the 8-site cube as the largest cluster, (9) for the bulk are:

$$\begin{aligned} &a_8 = 1 \\ &a_4 + 2\,a_8 = 1 \\ &a_2 + 4\,a_4 + 4\,a_8 = 1 \\ &a_1 + 6\,a_2 + 12\,a_4 + 8\,a_8 = 1 \end{aligned} \tag{10}$$

where n = 8, 4, 2, 1 refer to the cube, square, nearest-neighbor pair and point clusters, respectively. Thus, in the bulk, we obtain $a_8 = 1$, $a_4 = -1$, $a_2 = 1$, and $a_1 = -1$.

For the case where the translational symmetry of the lattice is broken by a surface, we must distinguish between the coefficients a_n of clusters in the bulk from the coefficients a_n^o of clusters located in the surface region (i.e. clusters that have all or some of their points on the surface). In the case of a (100) surface in the cube approximation, the coefficients for all subclusters vanish and $a_8^o = 1$. The coefficients for the fcc lattice with a (100) surface in the tetrahedron approximation can be found in Reference 13. Further details on how to solve the self-consistent equation (8) may also be found in Reference 13.

A comparison of the magnetizations as a function of temperature in the Bragg-Williams (BW) and cube (C) approximations is shown in Figure 1(a). The temperature scale is normalized to the corresponding critical temperatures T_b; equal to $6J_b$ and 4.581 J_b for the BW and C, respectively. As expected, the values of the magnetization in the C approximation are significantly larger than in BW. It is worth noticing that the best estimate of the value for T_b is 4.5132 [14]. Figure 1(b) contains on the same scale the squared magnetizations. In contrast to the BW-results, one observes that the C-results are linear only over a small range of temperatures near T_b.

The phase diagram for the (100) simple-cubic structure, showing the regions where the surface orders at the bulk temperature T_b and where $T_s > T_b$ is shown in Figure 2. The value for the surface critical exchange constant is 1.47 J_b. This value (in units of J_b) should be compared with that obtained by means of other techniques: 1.25 in the Bragg-Williams approximaion [5]; 1.307 by cumulant expansion and 1.357 by means of the renormalization group [3]; 1.3068 and 1.3297 in effective-field theories [15,16]; 1.50 ± 0.03 by Monte Carlo calculations [2]; 1.6 ± 0.1 by series expansion [7]; and 1.569 in real-space renormalization group [4].

The temperature dependence of the bulk and surface magnetizations for three different cases are plotted in Figure 3. Figures 3(a) and 3(b) contain cases where $J_s > J_c$ (0.5 J_b and J_b). A case for $T_s > T_b$ (extraordinary transition) is shown in Figure 3(c).

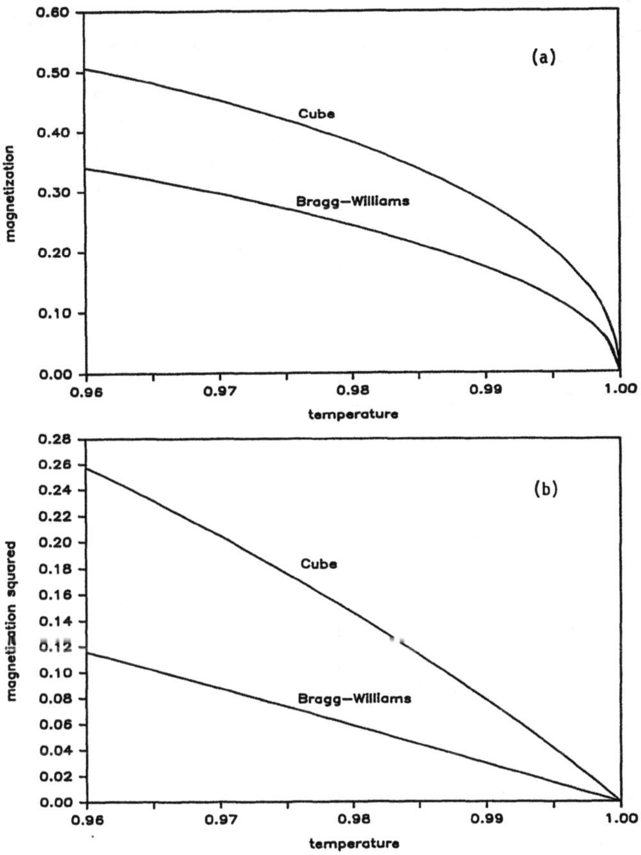

Figure 1. (a) Temperature dependence of the bulk magnetization in the cube and Bragg-Williams approximations in a simple-cubic lattice. (b) The bulk magnetization squared as a function of the temperature in the cube and Bragg-Williams approximations.

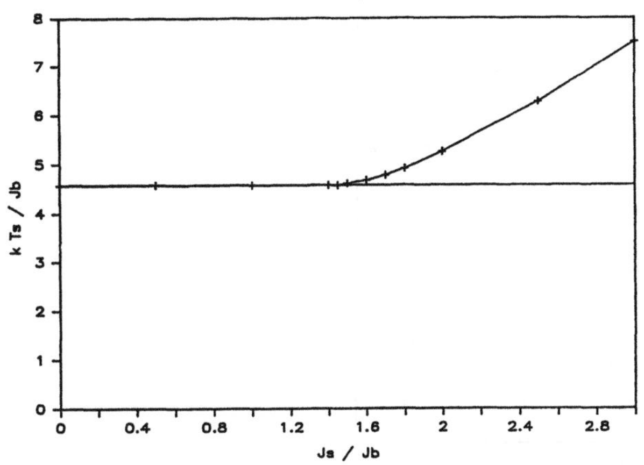

Figure 2. Phase diagram of kT_s/J_b vs. J_s/J_b as obtained in the cube approximation for the (100)-simple-cubic surface. For $J_s/J_b > 1.47$ the surface orders at $T_s > T_b$.

We have carried out a similar analysis for the (100) surface of a face-centered cubic crystal. The approximation used in this case is the one that has a tetrahedron as a basic cluster. The phase diagram obtained is shown in Figure 4 and the value for J_c is 2.25. Despite the fact that this crystal structure is more relevant from the experimental point of view, the only other estimate of J_c has been obtained in the BW approximation (2.0 J_b).

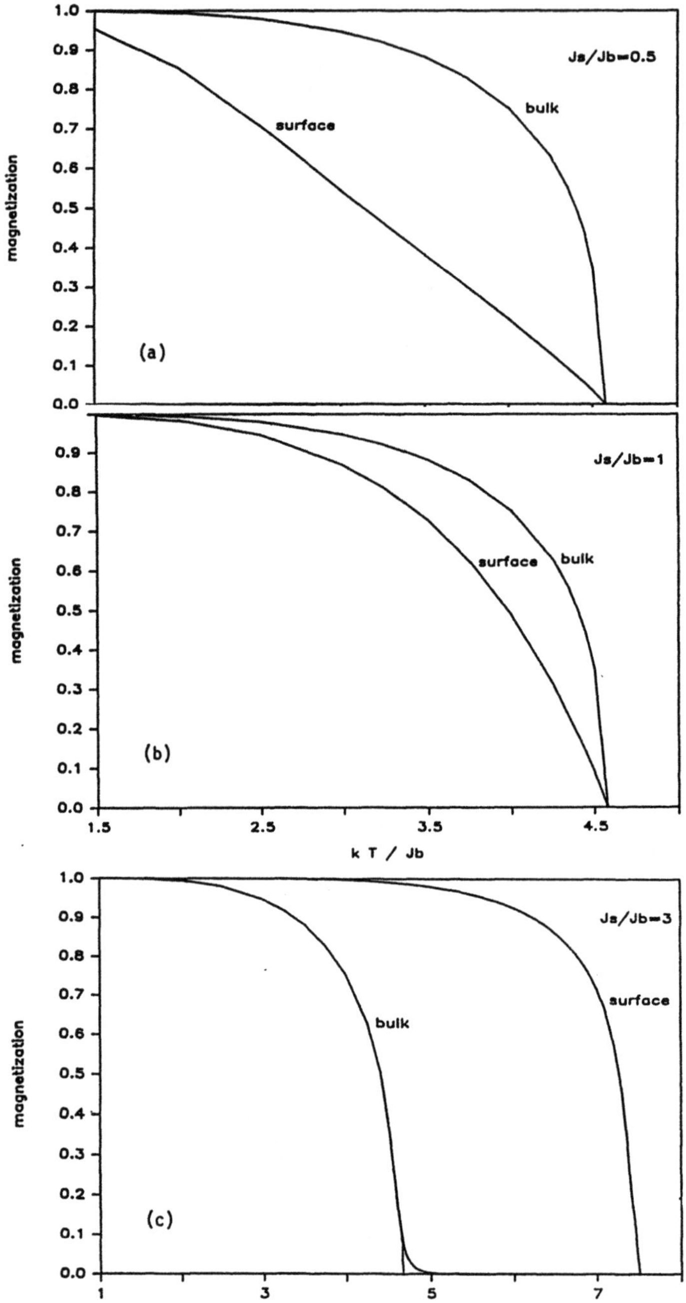

Figure 3. Surface and bulk magnetizations as a function of temperature for the (100) surface of a simple-cubic lattice. (a), (b), and (c) correspond to J_s/J_b = 0.5, 1.0, and 3.0, respectively.

Lastly, in Figure 5, we analyze the ratio of surface-to-bulk magnetization near the critical temperature. As seen in the figure, for values of $\tau = 1 - T/T_b$ smaller than 10^{-2}, the classical behavior, corresponding to zero slope, dominates. On the other hand, it is evident that for intermediate values of τ the slope is very close to the value predicted by the Monte Carlo calculations [2] for the difference between surface ($\beta_s \approx 0.78$) and bulk ($\beta_b \approx 0.3125$) critical exponents.

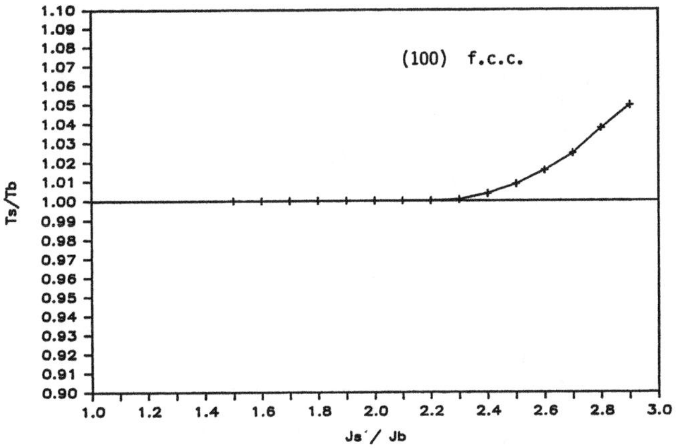

Figure 4. Phase diagram of kT_s/J_b vs. J_s/J_b as obtained in the tetrahedron approximations for the (100) fcc surface, the critical temperature value is $J_s/J_b = 2.25$.

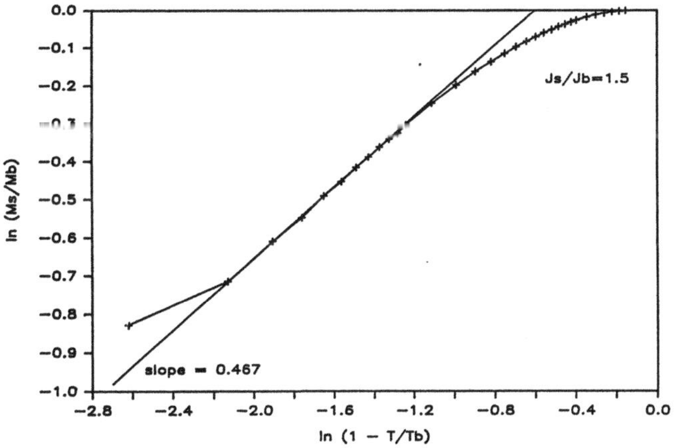

Figure 5. Temperature dependence of ln (M_s/M_b) for the (100) fcc surface as obtained in the tetrahedron approximation near the critical temperature. These results were obtained for $J_s/J_b = 1.5$.

Thus, our results suggest that the surface critical exponent for the fcc lattice falls in the universality class as that for the simple cubic lattice. However, a more extensive investigation is necessary before a definitive conclusion may be drawn.

In conclusion, we have shown that the cluster method may be used to estimate accurately the surface critical temperature of semi-infinite systems. We have applied the technique to the study of the important case of fcc lattices, for which only molecular field estimates of the J_c were available. In addition, our calculations suggest a limiting value of the critical exponent for the surface magnetization of 0.78, as found by Monte Carlo simulations in simple cubic lattices. Since the cluster method may be easily extended to include both more distant neighbor as well as many-body interactions, it should provide a valuable tool for the interpretation of experimental results.

Acknowledgments

This work was partially supported in the U.S. by the National Science Foundation through Grants Nos. DMR-8206195 and INT-8409776 and in Mexico by the Consejo Nacional de Ciencia y Tecnología through Grant PCCBBEU-022007 and by COSNET-SEP.

References

1. K. Binder, in Phase Transitions and Critical Phenomena, ed. by C. Domb and J. L. Lebowitz (Academic, New York, 1983), Volume 8, Chapter 1.
2. K. Binder and D. P. Landau, Phys. Rev. Lett. 52, 318 (1984).
3. S. Dietrich and E. Eisenriegler, Phys. Rev. B 27, 2937 (1983).
4. D. F. Sarmento and C. Tsallis, J. Phys. C 18, 2777 (1985).
5. D. C. Mills, Phys. Rev. B 3, 3887 (1971).
6. F. Aguilera-Granja, J. L. Morán-López, and J. Urías, Phys. Rev. B 28, 3909 (1983).
7. F. Aguilera-Granja and J. L. Morán-López, Phys. Rev. B 31, 7146 (1985).
8. R. Kikuchi, Phys. Rev. 81, 998 (1951).
9. J. M. Sánchez, D. de Fontaine, and W. Teitler, Phys. Rev. B 26, 1465 (1982).
10. J. M. Sánchez and D. de Fontaine, Phys. Rev. B 21, 216 (1980).
11. J. M. Sánchez and D. de Fontaine, Phys. Rev. B 17, 2926 (1978).
12. J. M. Sánchez, F. Ducastelle, and D. Gratias, Physica 128A, 334 (1984).
13. J. M. Sánchez and J. L. Morán-López, Phys. Rev. B 32, 3534 (1985).
14. D. de Fontaine, in Solid State Physics, ed. by H. Ehrenreich, F. Seitz and D. Turnbull (Academic, New York, 1979), Volume 34, p. 74).
15. E. F. Sarmento, I. Tamura, and T. Kaneyoshi, Z. Phys. B 54, 241 (1984).
16. T. Kaneyoshi, I. Tamura, and E. F. Sarmento, Phys. Rev. B 28, 6491 (1983).
17. K. Binder and P. C. Hohenberg, Phys. Rev. B 6, 3461 (1972).

Part IV

Thin Films

Very Thin Ferromagnetic Single Crystal Films

U. Gradmann

Physikalisches Institut der Technischen Universität Clausthal,
D-3392 Clausthal-Zellerfeld, Fed. Rep. of Germany

Very thin ferromagnetic single crystal films can be used for the experimental analysis of magnetic size effects and of surface magnetism. The main emphasis of this report is on experiments with Ni(111) and Fe(110) films, performed in three modern instruments combining both preparation and testing the films in UHV and magnetic measurements in situ with three magnetic probes: A Torsion Oscillation Magnetometer (TOM) detecting less than 10^{-2} monolayers Ni, Spin Polarized Low Energy Electron Diffraction (SPLEED), and monolayer probe Conversion Electron Mössbauer Spectroscopy (CEMS). Results are reported for magnetic size effects, for changes of interface magnetization in Ni(111) caused by Cu- or Pd-coatings, for the enhancement of J_s in a free Fe(110) surface, for Friedel-type oscillations of B_{hf} near a free Fe(110) surface, and for magnetic surface anisotropies in Ni(111) and Fe(110).

Ferromagnetic films consisting of few parallel atomic layers have been used for many years as models for theoretical analysis of magnetic size effects [1-3]. Recently they were used, too, as a powerful tool in the modern theoretical studies of surface and interface magnetism, mainly by means of band calculations of thin films [4-6]; for a review see, for instance Reference 7 and theoretical contributions in this volume.

The aim of the present paper is to discuss an analogous situation in experiment: Atomically flat ferromagnetic single crystal films, prepared by epitaxial growth on nonmagnetic single crystal substrates, which had been used previously for experimental analysis of magnetic size effects [8-10], have been used in our group also as a powerful tool for experimental studies of surface and interface magnetism. For a review of earlier work of this type refer to Reference 10. In the present review, we concentrate on a modern approach to this type of analysis, based on a new generation of instruments. They combine, in one UHV system, both epitaxial preparation and structural testing of the films and in situ magnetic analysis. We use three instruments of this type, each equipped with one of the following magnetic probes: a TOM, capable to detect less than 10^{-2} monolayers of Ni(111) [11], SPLEED [12], and a monolayer probe CEMS [13,14].

In the following sections, we discuss some results on (1) magnetic size effects, (2) magnetic interaction with nonmagnetic coatings, (3) local structure of magnetic order, and (4) magnetic surface anisotropies.

1. Magnetic Size Effects.

Finite size in one dimension results, independently of the special chemical state of the surface, in the following characteristics of magnetic order in thin films:

1.1 Decrease of the Curie Temperature $T_c(D)$ with Decreasing Thickness.

A decrease of Curie temperature $T_c(D)$ with decreasing number of atomic layers, D, was predicted first by mean-field arguments [3]. Later on, high temperature series expansions of the Heisenberg model [15] resulted in the prediction of a power law

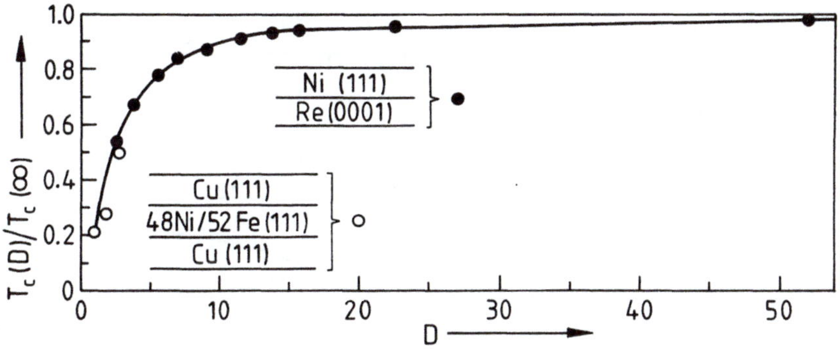

Figure 1. Curie temperatures $T_c(D)$ as a function of the number D of atomic layers, normalized to $T_c(\infty)$, for 0.48Ni/0.52Fe(111) on Cu(111) and Ni(111) on Re(0001) (from Reference 16).

$$(T_c(\infty) - T_c(D)) / T_c(\infty) = C_0 \cdot D^{-\lambda} \tag{1}$$

with a critical exponent $\lambda = 1/\nu$. Using $\nu = 0.705$ for the correlation length exponent, this results in $\lambda = 1.42$. TOM experiments with Ni(111) films on Re(0001) [11,16] resulted in experimental values of $T_c(D)$, which were in good agreement with former work on NiFe(111) films deposited on Cu(111) [9], as shown in Figure 1. Curve-fitting with (1) results in an experimental exponent $\lambda = 1.27 \pm 0.2$, slightly lower than expected from theory.

This low value may be connected to the comparatively thin film ($D < 60$) used in the experiment, as theory applies to the large D limit. For comparison, previous measurements with polycrystalline Ni films with $D > 100$ [17] resulted in $\lambda = 1.33 \pm 0.13$.

Of special interest are ferromagnetic monolayers. A detailed analysis of monolayer magnetism has been given for Co(0001) on Cu(111) [18], resulting in $T_c(1) = 0.31\ T_c(\infty)$, and a linear decrease of the magnetization J_s with decreasing T in a wide temperature range.

1.2 Changed Thermal Decrease of the Magnetization.

The spontaneous magnetization $J_s(T,D)$ becomes a function of both the temperature T and number D of atomic layers in the ferromagnetic film. The general dependence of $J_s(T,D)$ for Ni(111) films on Re(0001), measured for $T > 1/2\ T_c(\infty)$ [11], fits well to former measurements with 0.48Ni/0.52Fe(111) films [8,9] for $T < 1/2\ T_c(\infty)$, with appropriate normalization of T to $T_c(\infty)$ and of J_s to $J_s(0,\infty)$ [16]. For low temperatures, results are in fine agreement with the spin-wave theory of thin films [2,9,19]. The linear decrease of the monolayer magnetization $J_s(T,1)$, expected from spin-wave theory at low temperatures, persists up to the Curie temperature $T_c(1)$ of the monolayer [9,18]. A theoretical explanation of this simple result is not available.

2. Changes of the Magnetic Moment by Interaction with a Nonmagnetic Coating.

Any changes in the electronic structure of a magnetic surface caused by nonmagnetic coatings, gas adsorptions, or other reactions must result in a change of the magnetic moment per surface area. This can easily be tested using films, when the magnetic moment can be followed by magnetometry, as has been shown first for the interaction of a 0.60Ni/ 0.40Fe(111) surface with a Mn-coating [9,20,21], which results in a decrease of the moment by antiferromagnetic coupling of the first Mn-layer to the last NiFe-layer. These experiments were done by magnetometry in

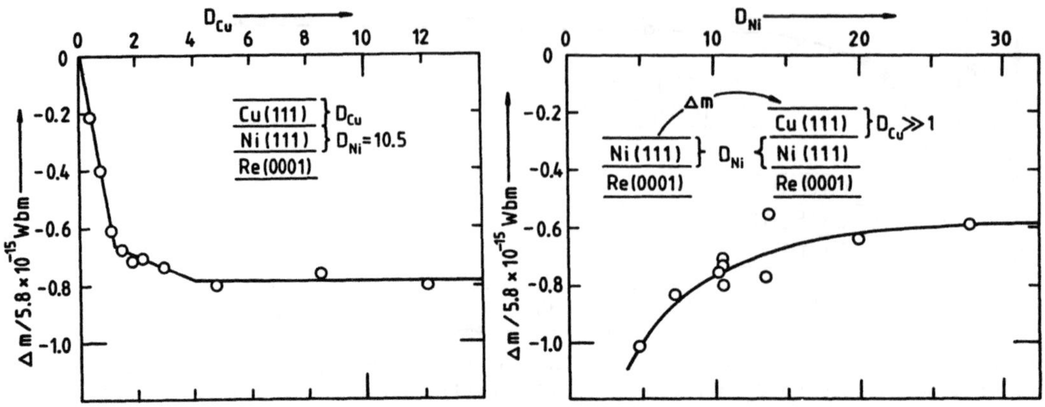

Figure 2. Change in the magnetic moment, Δm, for Ni(111) films, caused by a Cu(111) coating (in units of 5.8×10^{-15} Wb m, the moment of 1 monolayer of bulk Ni(111): (a) Δm as a function of D_{Cu} for a nickel film with $D_{Ni} = 10.5$; (b) saturation value of Δm, for $D_{Cu} \gg 1$, as a function of D_{Ni} (from [16]).

air and comparison of Mn-coated with Ag-coated films. Only recently changes of this type became accessible to theoretical calculation [6,22,30]. They can be tested now most effectively in the UHV-TOM, where we can follow the changes of magnetic film moment in situ during coating or reaction processes. This is shown in Figure 2 for the example of a Cu-coating on Ni(111) [23]. The decrease of magnetic moment caused by thick Cu on thick Ni, equivalent to 0.6 monolayers Ni(111) as shown in Figure 2(b), is in remarkably good agreement with the calculations of TERSOFF and FALICOV [6], who predicted a decrease by 0.61 monolayer equivalents for a coating with two monolayers Cu(111).

Coating of Ni(111) with Pd(111) results in an increase of the film magnetic moment by magnetization of the coating Pd(111) [24]. This incremental moment was calculated recently by J. MATHON [25] within a Ginzburg-Landau model, as shown in Figure 3.

3. Local Structure of Magnetic Order.

Two types of inhomogeneities of the magnetization are expected near surfaces and in thin films. Thermal excitations result in a monotonic decrease of magnetization

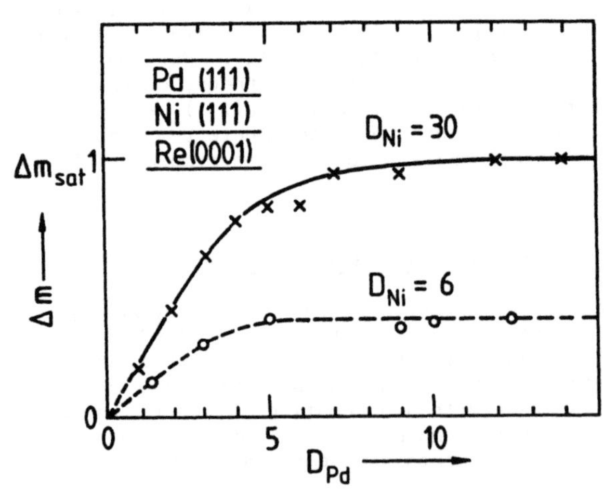

Figure 3. Incremental magnetic moment Δm of Ni(111) films of D_{Ni} atomic layers, caused by a coating Pd(111) film of D_{Pd} layers, in units of the saturation increase Δm_{sat} for $D_{Ni} = 30$, $D_{Pd} > 10$, which in turn corresponds to an additional moment of 0.62 μ_B/Pd-atom, when projected to one single interface layer. Experiments from [24] (x,o) are compared with theoretical results from [25] (——,- - -).

from the bulk to the surface; this has been discussed by means of molecular-field theory [3], by spin-wave theory [26], by Monte Carlo calculations [27], and in the critical range by renormalization group analysis [28]; for a review see Reference 29. This thermally induced decrease is superimposed to ground-state inhomogeneities of the magnetic order, which were calculated only recently by means of band theory [4,31,32]. Non-monotonic, oscillatory inhomogeneities were predicted [4,32], in contrast to the monotonic decrease induced thermally.

Both types of inhomogeneities can be analyzed by SPLEED, which was first applied to Ni(100) by FEDER et al. [33] (for a review of magnetic SPLEED see, for instance, Reference 34) or by Mössbauer analysis using ^{57}Fe probe layers. Of special note is the pioneering work of WALKER et al. [35] on Ag-coated Fe(110) surfaces. The merit of SPLEED is its general applicability to any free surface; the merit of Mössbauer spectroscopy is its extreme accuracy and its applicability to interfaces. Mössbauer spectroscopy is restricted to Fe surfaces and to $T < 500$ K, because of ^{57}Fe-^{56}Fe-interdiffusion at higher temperatures [13]; it results in a local structure of the magnetic hyperfine field B_{hf}, which must be distinguished carefully near surfaces from the local structure of J_s; this fact has been stressed recently by OHNISHI et al. [32]. Modern theories, however, provide a link between B_{hf} and J_s [32].

Both methods have been applied in our group to Fe(110) films on W(110), based on previous studies of the epitaxial growth in this system [36]. The use of films has the merit of easy access to the clean Fe surface.

For SPLEED, the comparison of experimental exchange asymmetries A_{ex} in polarized electron scattering [12] with theoretical results of TAMURA et al. [37] resulted in the statement that the magnetization of the first monolayer in Fe(110) is enhanced

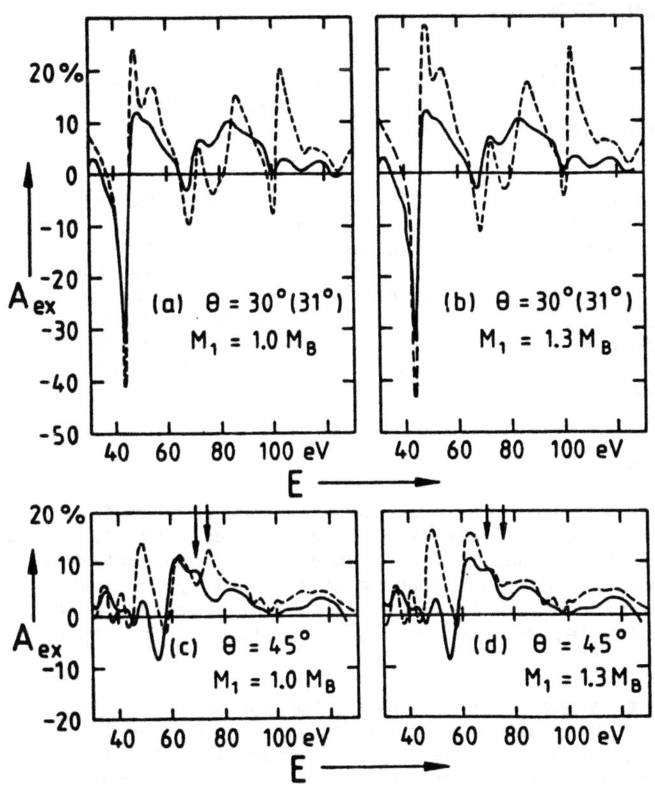

Figure 4. Exchange asymmetry A_{ex} as a function of kinetic energy E for specular reflection of spin-polarized electrons from Fe(110). Experiments for incident angles $\theta = 31°$ and 45°, respectively (full curves) from [12] are compared with calculations for $\theta = 30°$ and 45°, respectively (dotted curves) from Reference 37. For Figures 4(a) and (c) the calculations were done with a magnetization M_1 of the topmost layer given by M_1 = 1.0 M_B (bulk magnetization M_B). For Figures 4(b) and (d), M_1 = 1.3 M_B is assumed for the calculations (from Reference 38).

by roughly 30% [38], as seen in Figure 4. Refinements of both theoretical calculations and experiments, including the measurement of A_{ex} in nonspecular beams, should be available soon [39].

For Mössbauer spectroscopy, the use of CEMS in combination with the use of a sample as input window of a high luminosity electron spectrometer and with the use of Fe(110)-film samples, enabled the real monolayer resolution analysis of B_{hf} near surfaces, by incorporation of monolayer ^{57}Fe probes in Fe(110) films consisting otherwise of ^{56}Fe [13,14,40]. Near free Fe(110) surfaces, oscillatory inhomogeneities could be detected at room temperature, as seen in Figure 5.

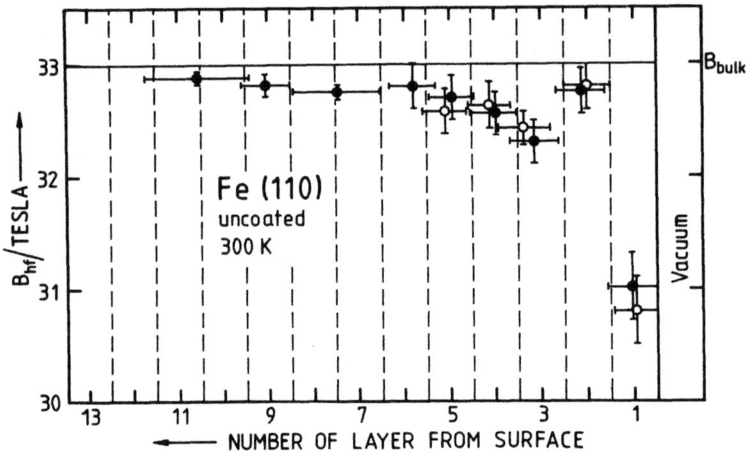

Figure 5. Magnetic hyperfine field B_{hf} near a free Fe(110) surface, at 300 K. Preparation of the 21-layers Fe(110) films on W(110) started at 570; it was lowered to 420 (in order to prevent interdiffusion) after 13 layers (●) or for the probe layer and all following layers (o) (from [40]).

For Ag-coated surfaces, these inhomogeneities are strongly reduced; however, the thermally induced decrease near the surface could be analyzed for this case, and good agreement with spin-wave predictions could be achieved [40], as shown in Figure 6. Using this thermally induced decrease, which should be the same for the free surface, the ground-state inhomogeneities of the free Fe(110) surface could be determined by extrapolation, as shown ine Figure 7. The amplitude of the B_{hf} oscillations is more than an order of magnitude lower than predicted for Fe(100) [4,31]; this may be connected with the more open character of Fe(100) as compared with the most densely packed Fe(110) surface.

4. Magnetic Surface Anisotropies.

As a result of the strong symmetry distortion caused by a surface or interface, strong magnetic surface (interface) anisotropies are expected; these were discussed first by L. NÉEL [41], who presented a quantitative connection between surface and strain anisotropies in a phenomenological pair-bonding model. Thirty-two years after Néel's pioneering work, a microscopic theory of surface anisotropy, beyond Néel's contribution, is not available, with the possible exception of one paper [42].

The technical importance of this phenomenon, however, for magnetic reversal properties in small particle and thin film systems, which was noted by Néel himself, should be emphasized again today in the light of recent trends towards miniaturization of magnetic recording devices and of magnetic superlattices.

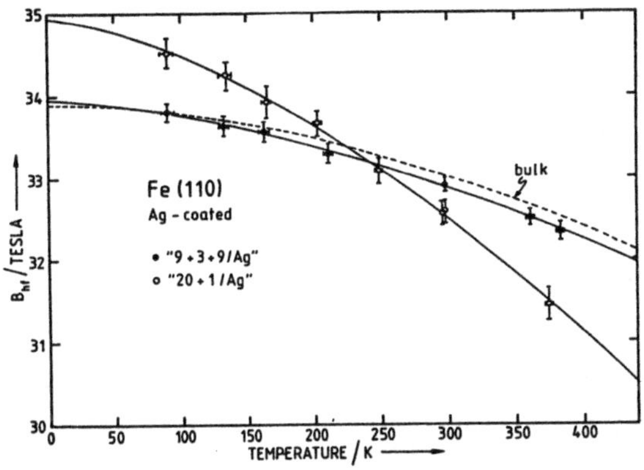

Figure 6. Magnetic hyperfine field B_{hf} in the center (●) and in the first monolayer (o) of a 21-layer Fe(110) film on W(110), coated with Ag, as a function of temperature. Lines give a spin-wave fit (from Reference 40).

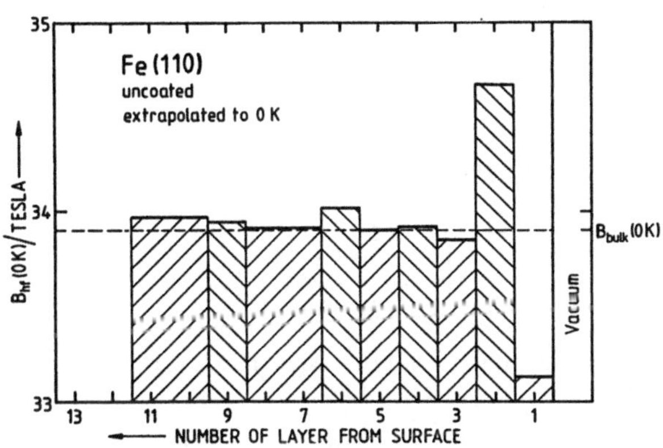

Figure 7. Magnetic hyperfine field B_{hf} at T = 0 K near a free Fe(110 surface (from Reference 40).

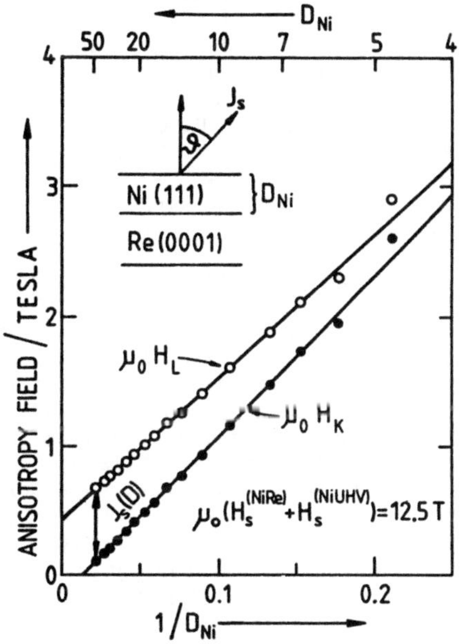

Figure 8. Anisotropy fields $\mu_0 H_L$ and $\mu_0 H_K = \mu_0 H_L - J_s$, for Ni(111) films on Re(0001), as a function of $1/D_{Ni}$.

The analysis of anisotropies in well-defined thin-film structures provides a powerful tool for the experimental analysis of magnetic surface anisotropies; for a review see Reference 43. While continuing previous work on NiFe(111) film with magnetometry in air [9], very clear evidence of magnetic surface anisotropies in Ni(111) surfaces and interfaces came from magnetometry of Ni(111) films in the UHV-TOM. Results are shown in Figure 8 in terms of an anisotropy field $H_L = 2L/J_s$, where $F_L = L \cos^2 \Theta$ is the anisotropic magnetic contribution to the free energy of the film, and depends on the angle Θ between J_s and the film normal. The linear increase of H_L and $(H_L - J_s)$, with increasing $1/D_{Ni}$ gives clear evidence for magnetic surface anisotropies in both surfaces, the sum of which can be determined from the slope. Note that the thick film shape anisotropy is of the order 0.6 Tesla, so that the increase of H_L indicates a strong enhancement of anisotropy with decreasing D: Surface anisotropies in Ni(111) are easy plane anisotropies. A phenomenological combination of experiments with various coating materials shows that the clean surface has an anisotropy field of 10 Tesla, whereas the metal-coated surfaces show only 4 Tesla [43,44]. For the less symmetric Fe(110) surface, in-plane surface anisotropies could also be determined using CEMS and SPLEED with Fe(110) films on W(110) [45].

References

1. M. J. Klein and R. S. Smith, Phys. Rev. 81, 378 (1951).
2. W. Döring, Z. Naturf. 16a, 1008, 1146 (1961).
3. L. Valenta, Czech. J. Phys. 7, 127 (1957).
4. C. S. Wang and A. J. Freeman, Phys. Rev. B 21, 4585 (1980).
5. O. Jepsen, J. Madsen, and O. K. Andersen, J. Magn. Magn. Mat. 15-18, 867 (1980).
6. J. Tersoff and L. M. Falicov, Phys. Rev. B 26, 6186 (1982).
7. A. J. Freeman, J. Magn. Magn. Mat. 35, 31 (1983).
8. U. Gradmann and J. Müller, phys. stat. sol. 27, 313 (1968).
9. U. Gradmann, Appl. Phys. 3, 161 (1974).
10. U. Gradmann, J. Magn. Magn. Mat. 6, 173 (1977).
11. R. Bergholz and U. Gradmann, J. Magn. Magn. Mat. 45, 389 (1984).
12. G. Waller and U. Gradmann, Phys. Rev. B 26, 6330 (1982).
13. J. Korecki and U. Gradmann, Phys. Rev. Lett. 55, 2491 (1985).
14. J. Korecki and U. Gradmann, Hyperfine Interactions, in print.
15. C. A. T. Allan, Phys. Rev. B 1, 352 (1970).
16. U. Gradmann, R. Bergholz, and E. Bergter, Thin Sol. Films 126, 107 (1985).
17. H. Lutz, J. P. Gunton, H. K. Schurmann, J. Crow, and T. Mihalisin, Solid State Comm. 14, 1075 (1974).
18. U. Gradmann and J. Müller, Czech. J. Phys. B 21, 553 (1971).
19. R. J. Jelitto, Z. Naturf. 19a, 1567, 1580 (1964).
20. U. Gradmann and K. Salewski, Physica 86-88B, 1397 (1977).
21. U. Gradmann and K. Salewski, phys. stat. sol. (a) 39, 41 (1977).
22. C. L. Fu, A. J. Freeman, and T. Oguchi, Phys. Rev. Lett. 54, 2700 (1985).
23. E. Bergter, U. Gradmann, and R. Bergholz, Solid State Comm. 53, 565 (1985).
24. U. Gradmann and R. Bergholz, Phys. Rev. Lett. 52, 771 (1984).
25. J. Mathon, to be published.
26. D. L. Mills and A. A. Maradudin, J. Phys. Chem. Solids 28, 1855 (1967).
27. K. Binder and P. C. Hohenberg, Phys. Rev. B 9, 2194 (1974).
28. H. W. Diehl and S. Dietrich, Phys. Lett. 80A, 408 (1980).
29. K. Binder, in Polarized Electrons in Surface Physics, ed. by R. Feder (World Scientific, Singapore, 1985).
30. D. Wang, A. J. Freeman, and H. Krakauer, Phys. Rev. B 24, 1126 (1981).
31. O. Jepsen, J. Madsen, and O. K. Andersen, Phys. Rev. B 26, 2790 (1982).
32. S. Ohnishi, M. Weinert, and A. J. Freeman, Phys. Rev. B 30, 36 (1984).
33. R. Feder, S. F. Alvarado, E. Tamura, and E. Kisker, Surf. Sci. 127, 83 (1983).
34. U. Gradmann and S. F. Alvarado, in Polarized Electrons in Surface Physics, ed. by R. Feder (World Scientific, Singapore, 1985).
35. J. C. Walker, R. Droste, G. Stern, and J. Tyson, J. Appl. Phys. 55, 2500 (1984).
36. U. Gradmann and G. Waller, Surf. Sci. 116, 539 (1982).
37. E. Tamura and R. Feder, Solid State Comm. 44, 1101 (1982).
38. U. Gradmann, G. Waller, R. Feder, and E. Tamura, J. Magn. Magn. Mat. 31-34, 883 (1983).
39. E. Tamura, R. Feder, G. Waller, and U. Gradmann, to be published.
40. J. Korecki and U. Gradmann, to be published.
41. L. Néel, Compt. Rend. 237, 1468 (1953).
42. H. Takayama, K. P. Bohnen, and P. Fulde, Phys. Rev. B 14, 2287 (1976).
43. U. Gradmann, J. Magn. Magn. Mat., to be published.
44. U. Gradmann, R. Bergholz, and E. Bergter, IEEE Trans. Magn. 20, 1840 (1984).
45. U. Gradmann, J. Korecki, and G. Waller, Appl. Phys. A 39, 1 (1986).

Part V

Small Particles and Clusters

Fine Particle Magnetism

E.P. Wohlfarth

Department of Mathematics, Imperial College, London SW7 2BZ, England

1. Introduction.

This subject combines the two aspects of magnetism, i.e., the scientific and the technological. Scientific aspects involve the ideas of:

Classical magnetism, exchange;
Domains, critical size for single domains;
Micromagnetics;
Hysteresis curves;
Remanence curves;
Irreversible processes;
Thermal relaxation effects;
Global optimization via anhysteresis, etc.

Technological aspects include the properties and uses of these magnetic materials:

Spin glasses (!);
Rocks;
Ferrofluids;
Magnetic recording media;
Permanent magnets;
Living organisms;
Catalysts, etc.

Magnetic information technology (MINT) requires a particularly intensive study of fine particle magnetism: Silicon Valley now produces more dollar value of magnetic components than of semiconductors.

2. Domain Walls; Critical Size for Single Domains.

For the Heisenberg model the exchange energy is given by

$$E_x = -2J\,\vec{S}_i \cdot \vec{S}_j = -2J\,S^2 \cos^2\phi = \text{const.} + J\,S^2\,\phi^2\ , \text{ approx.,}$$

where ϕ is the angle between two neighboring spins. If a change of ϕ by π takes place between neighbors, $E_x = J\,S^2\,\pi^2$, approximately. If a gradual change takes place by π over a row of N atoms, $E_x = J\,S^2\,(\pi/N)^2\,N = J\,S^2\,\pi^2\,/\,N$, much less. Domain walls thus have a finite width δ, determined by E_x and the anisotropy energy E_K. Let a be the atomic separation, then $\delta = Na$ and $E_K = K_1\,\delta$ per unit wall area, where K_1 is the (crystal) anisotropy constant. The total energy per unit wall area is

$$E = E_x + E_K = J\,S^2\,\pi^2\,/\,\delta\,a + K_1\,\delta,$$

which has a minimum if $\delta = \delta_c = (J\,S^2\,\pi^2\,/\,K_1\,a)^{1/2}$.

If a fine particle has a size $< \delta_c$, it will be roughly a single domain, i.e., δ_c is roughly the critical single domain size. For an elongated particle (or acicular particle) the anisotropy is that of shape and $K_1 \to N_b M_S^2$, where N_b is the demagnetizing factor in the short direction. Hence

$$\delta_c \equiv b_c = (J S^2 \pi^2 / a N_b M_S^2)^{1/2} .$$

A better theory (micromagnetics) replaces π^2 by 4.32π for a cylindrical particle. Values of b_c are typically a few 100Å. Such single domain particles reverse their magnetization by rotational processes against the anisotropy, e.g. of shape, leading to magnetic hardness and many applications, as discussed later.

3. Hysteresis Curves of Single Domains.

For a single domain of ellipsoidal shape and semiaxes a, b the total energy per unit volume is

$$E = (1/2) M_S^2 (N_a \cos^2\psi + N_b \sin^2\psi) - H M \cos\phi ,$$

where Θ is the angle between the applied field H and the long axis a (x direction), ψ the angle between this direction and the magnetization M_S and $\phi = \Theta + \psi$. Minimizing or maximizing E gives

$$(1/2) \sin^2(\phi - \Theta) + h \sin\phi = 0, \qquad h = H / (N_b - N_a) M_S ,$$

giving $\psi = \psi(\Theta,h)$. Maxima, minima, and points of inflection lead to phase transitions at a field H with components H_x, H_y (y perpendicular to x) lying on the famous asteroid,

$$H_x^{2/3} + H_y^{2/3} = H_c^{2/3}(0) ,$$

where $H_c(0)$ is the coercive force of the resultant hysteresis loop with $\Theta = 0$, a square loop. For $\Theta = \pi/2$, the loop degenerates into a straight line and intermediate loops are obtained. The angular dependence $H_c(\Theta)$ thus computed may not occur if the reversals of the magnetization are incoherent as a result of micromagnetic principles which imply the lowering of the energy thereby. For the above case of coherent reversals an average hysteresis loop is obtained by averaging Θ-loops, and for random orientations the resulting mean remanence $M_r/M_S = 0.5$ and mean coercivity $H_c = 0.479 (N_b - N_a) M_S$ are calculated. The latter gives about 5000 oersted for Fe particles and about 2500 oersted for (aligned, Θ=0) γFe_2O_3 particles. Such high values are not usually observed, for the following reasons:

(1) Particle interactions, see later;
(2) Thermal fluctuation effects, see later;
(3) Imperfect surface conditions, dendrites, etc.;
(4) Mixed anisotropies (e.g. crystal + shape);
(5) Incoherent magnetization reversals, etc.

The latter may be "fanning" if the particles are chains of spheres (as in magnetotactic bacteria, see later), where neighboring spheres reverse in opposite directions, or "curling" in long cylinders (imagine a bundle of wires twisted to represent the spin directions, thus reducing the surface pole energy). This reversal mode modifies $H_c(\Theta)$ and reduces, as does fanning, the mean coercive force. For curling,

$$H_c = H_c(\text{coherent}) \qquad \text{for } b < b_c ,$$

$$H_c = H_c(\text{coherent}) (b_c/b)^2 \qquad \text{for } b > b_c .$$

4. Remanence Curves.

These are essential to understand technological applications but are also scientifically fascinating as they model irreversible processes. The so-called static or isothermal remanence curve (IRM) is obtained from a state of demagnetization by applying a DC field H and then reducing it to zero, thus catching the irreversibly attained remanence $M_r(H)$. For a previously saturated assembly of particles application and reduction of a reverse DC field H gives the DC demagnetization remanence curve $M_d(H)$. A similar demagnetization with an AC field of initial amplitude H gives the AC demagnetization remanence curve $M'_d(H)$. Finally the anhysteretic remanence curve (ARM) is obtained by applying a DC field H and a large AC field whose amplitude is slowly reduced to zero, followed by switching off H, giving $M_{ar}(H)$; this unusually important curve is discussed later.

If there are no interactions between the particles, then

$$M_d(H) = M_r(\infty) - 2\,M_r(H)\ ,$$

$$M'_d(H) = M_r(\infty) - M_r(H)\ ,$$

$$M_{ar}(H) = M_r(\infty)\ .$$

Hence such interactions may be assessed by testing these relations.

(1) Measure the remanence coercivity either from $M_d(H_r) = 0$ or from $M_r(H'_r) = M_r(\infty) / 2$ where $H_r = H'_r$ only without interactions. An interaction field factor (IFF) was defined [1] as $(H'_r - H_r) / H_c$ and measured for recording media, as shown below (Figure 1).

(2) The anhysteretic susceptibility $dM_{ar}(H)/dH = \chi_{ar}$ is only finite in the presence of interactions which may be assessed by defining an internal demagnetization coefficient $N_i = \chi_{ar}^{-1}$.

5. Interactions.

These are central to the whole subject, difficult to treat analytically, amenable to computer simulation (see later) and to measurement as just stated. Figure 1 [1] gives the dependence of the coercive force of two different samples, D and P, of γFe_2O_3 powders on the interaction effect as measured by IFF; increased dilution increases H_c towards extrapolated values given by b/b_c, i.e., by the exchange integral and the size of the particles. The reader is invited to suggest experiments elsewhere which give such simple assessments of classical and quantum mechanical many-body effects!

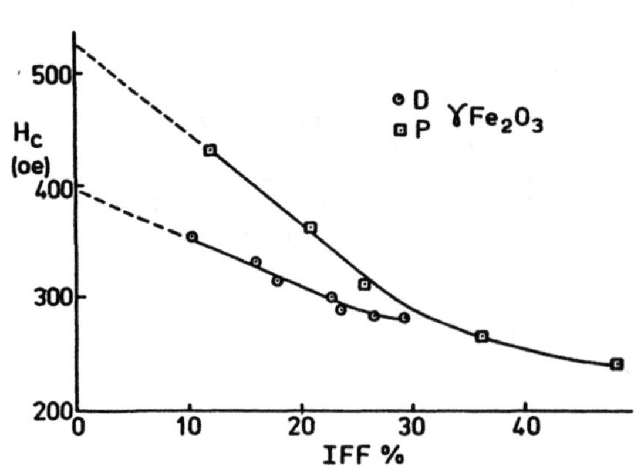

Figure 1. Dependence of coercive force on packing [1].

On dilution H_c may not always increase. Thus a single isotropic sphere has $H_c = 0$, but a chain of spheres, a finite coercive force due to the resultant "interaction anisotropy." We have obtained such a systematic trend more generally by the computer simulations discussed below.

6. Thermal Fluctuations and Time Effects.

These are again of both scientific and practical importance. Thus the thermal stability of permanent magnets and magnetic recording media is involved. Again, rocks (and spin glasses) attain thermoremanence by a gradual stabilization of thermal fluctuations (TRM). The relaxation phenomena are well defined and a consideration of the relevant rate equation shows that the magnetization of a particle depends on time through the relation

$$M(t) = S \exp(-t/\tau) .$$

Here the relaxation time τ is given by the barrier hopping process, the barrier height being defined by the anisotropy energy K (crystal, shape, ...), such that

$$\tau = \tau_0 \exp(vK/kT) ,$$

where τ_0 ~ inverse Larmor frequency and v the particle volume. A distribution of barrier heights due to variations of K, v, ... leads to a relation, usually applicable over extended time intervals, of the form

$$M(t) = M(0) - S \ln t ,$$

where S is discussed below. Other M(t) laws seem to be rather more ephermeral.

Applied field H. Here

$$\tau = \tau_0 \exp[\frac{vK}{kT} (1 - \frac{H}{H_K})^2] , \quad H_K \sim K/M_S .$$

Blocking temperature. If the relaxation time τ is short, the magnetization in the particle is thermally unstable and it is called superparamagnetic. Stabilization of this phenomenon occurs at a blocking temperature T_B given from the above by

$$v K / k T_B = - \ln(\nu\tau_0) ,$$

where ν is the frequency of the (AC) measurement. For a DC measurement in a time of 100 sec, the right-hand side of this equation is approximately equal to 25. Both the field dependence of T_B (Almeida-Thouless) and its frequency dependence are well known phenomena in spin glasses and explicable in fine particle terms. The frequency dependence can, however, often not be explained in terms of physical parameters due to particle interactions. These replace the formula for the relaxation time by the Vogel-Fulcher law

$$\tau = \tau_0 \exp[v K / k (T-T_0)] ,$$

applicable over extended ranges of temperature T and a universal law in solid-state physics and chemistry. Here T_0 is yet another measure of particle interactions.

Coefficient S. This important coefficient is given semiempirically by

$$S = \chi_{irr} H_f .$$

Here $\chi_{irr} = dM_r(H)/dH$ is the irreversible susceptibility appropriate to the barrier hopping process involved. The "fluctuation field" H_f is given by

$$v M_S H_f = k T ,$$

a formula useful for analyzing time-dependent data. The volume is more generally (i.e. beyond fine particle magnetism) that involved in a single activation process ("activation volume"). It depends on interaction effects so as to give a physical basis of the Vogel-Fulcher law [2].

7. Fine Particle Magnetism and Global Optimization.

This work [3] represents an intellectual jump from technology to numerical mathematics. Consider the anhysteretic remanent magnetization (ARM). As indicated above, a DC field H_{DC} plus an AC field is applied to the specimen, the latter having its amplitude decreased by an AC field decrement per cycle DH_{AC}. The resulting final remanence was simulated by the Monte Carlo technique for a 3D lattice of single domain particles in dipolar interaction. The ARM passes through the origin ($M_{ar} = 0$, $H_{DC} = 0$) only if the assembly has reached a global minimum as a result of DH_{AC} being small enough. If $M_{ar} = \pm S$, $H_{DC} = 0$, as shown in Figure 2, the AC field decrement was too large and only a local minimum was obtained. Hence S is here a measure of the non-attainment of a globally optimized state. Figure 3 shows S as a function of DH_{AC}, showing that S = 0 only until this decrement enters the distribution of coercive forces. Another way of simulating such processes was also considered [3]. Here a DC field is again applied but the amplitude of the AC field is kept constant for N cycles before reducing it by a decrement DH_{AC}, etc. The instability parameter S may be computed as a function on N and gives again a

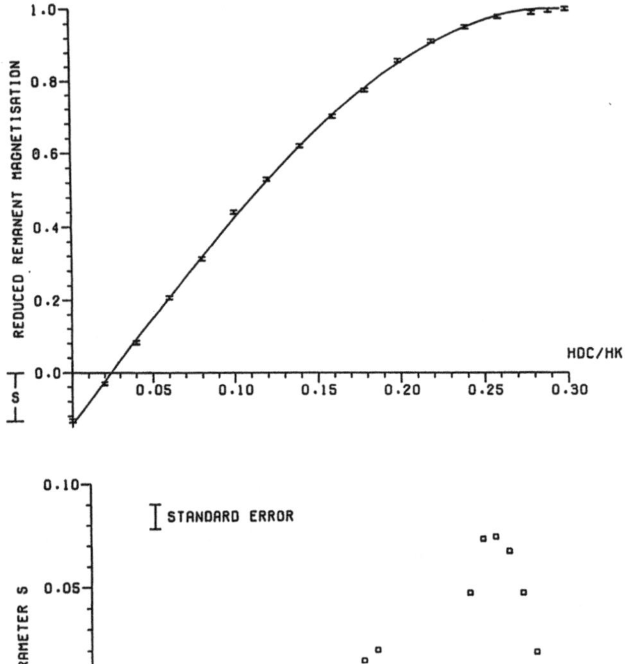

Figure 2. Unstable ARM curve, defining S [3].

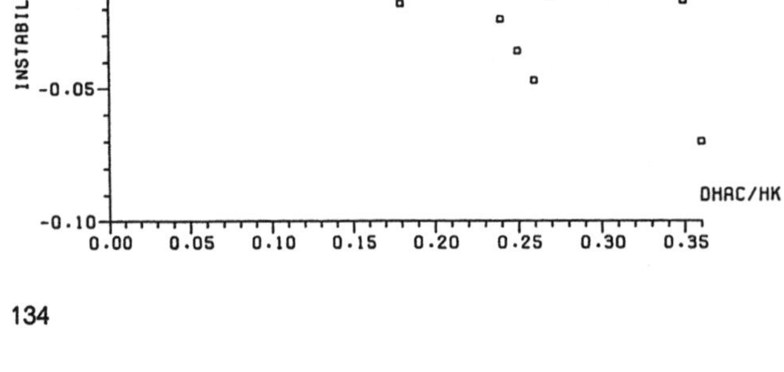

Figure 3. S as a function of the AC field decrement [3].

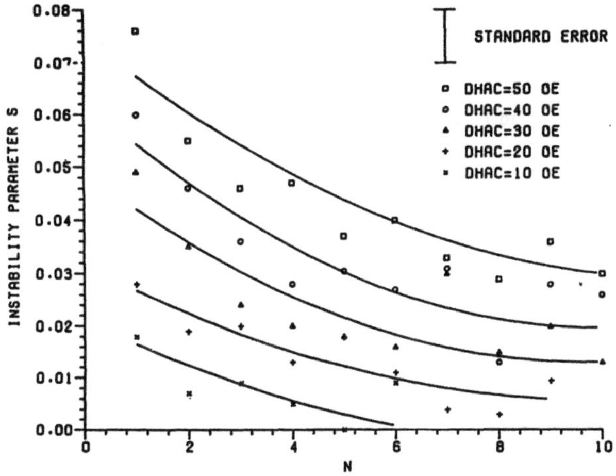

Figure 4. S as a function of the number of cycles in ACARM [3].

measure of global optimization. The results are shown in Figure 4. This process is called AC annealing remanent magnetization (ACARM) and is very analogous to Kirkpatrick's simulated annealing process. Our work is thus in a modern mainstream which is already extending into other branches of science (biology, ...). In Kirkpatrick's work the temperature is lowered, in ours the AC field amplitude, although temperature also enters, for help in barrier hopping.

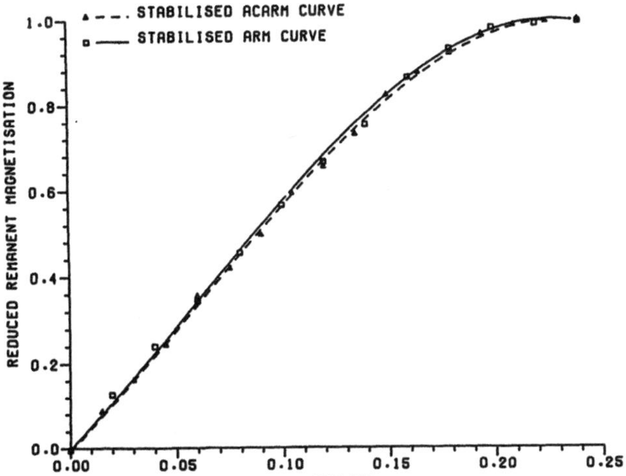

Figure 5. Globally optimized curves [3].

Figure 5 shows ARM and ACARM curves which have been globally optimized, with zero remanence at $H_{DC} = 0$. They have all the features of output curves for recording processes with AC bias, thus completing the technological part of this equation. The linear part corresponds to a distortion-free output signal with sensitivity given by χ_{ar}; distortions occur at higher signals ("disco effect") finally leading to saturation. Figure 6 shows the dependence of the growth of the remanent magnetization on the amplitude of the AC field during its gradual reduction. The anomalous maximum is again a sign of the non-attainment of a global minimum. However, this is an unusually simple example of a growth process which may again have relevance elsewhere (biology, ...?).

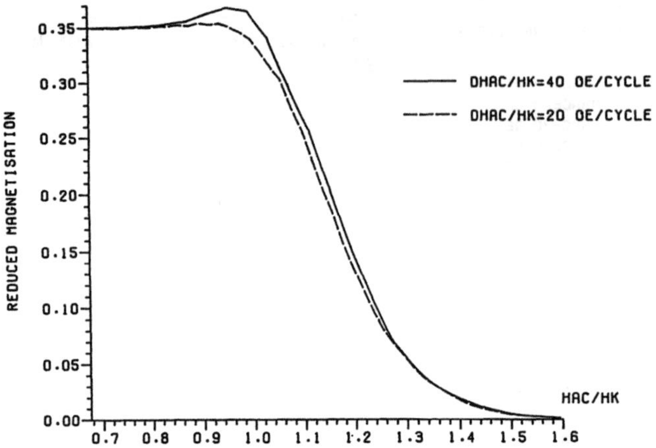

Figure 6. Growth of magnetization during AC field reduction [3].

8. Fine Particle Magnetic Materials.

(i) Spin Glasses.

The two approaches to the spin-glass problem (statistical-mechanical and fine-particle) are really basically similar in that distributions of energy minima occur in both. In the fine-particle approach the barriers are determined by the prevalent anisotropy. Spin-glass properties which can be explained in fine particle terms include:

Susceptibility peak at T_B;
Remanence below T_B, including TRM;
Time effects, e.g., as ℓnt;
Frequency effects via Vogel-Fulcher law;
Field effects, etc.

Spin glass materials which have been studied include Au-Fe, Cu-Mn, Pd-Mn, Gd-La-Al, $TiVO_3$, amorphous alloys, Eu-Sr-S, aluminosilicate glasses, etc. [4].

(ii) Rock Magnetism.

These are [5] materials containing fine ferromagnetic particles in organic matrices. Materials include Fe, Fe_3O_4, γFe_2O_3, αFe_2O_3, Fe_7S_8, Ti containing oxides, etc. Properties in fine particle magnetism which have been studied include:

Single domain sizes;
Superparamagnetism;
Coercive force distributions using $M'_d(H)$ curves;
TRM;
Chemical (or growth) remanence (CRM);
Viscosity and time effects;
$\chi(T)$ maxima.

The latter effect is shown in Figure 7 for a basaltic rock [6]. The sharp maximum is clearly indicative of blocking.

(iii) Magnetic Recording Media. [7,8]

Most fine particle concepts occur in the study of these media which are basic to magnetic information technology (MINT). Modern media include:

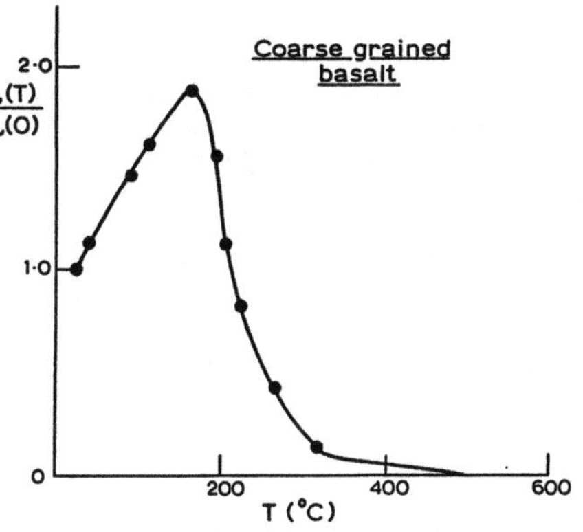

Figure 7. Susceptibility peak of a rock [6].

(a) Iron Oxide, γFe_2O_3.

Particles are produced by a reduction/oxidation cycle in the form of elongated particles about 0.5 μ long and of elongation about 7:1. Coercive forces H_c ~ 250-320 oersted are obtained and are adequate for most purposes. Alignment in a field at the wet stage of dispersion gives $M_r(\infty)$ / M_s about 0.8-0.9. Improvement of these properties continues; thus particles without internal cavities (NP particles) were produced in Japan, thus reducing demagnetization effects. Scientific studies showed that the magnetization reversals were "more coherent" and H_c values about 400 oersted resulted.

(b) Co Adsorbed Epitaxial Fe_2O_3.

Here the crystal anisotropy of the adsorbed surface layer adds to the shape anisotropy of the core to give H_c > 600 oersted.

(c) Isotropic Particles.

These are potentially useful for perpendicular recording, a mode where demagnetization is reduced and thus information density increased. Cobalt containing particles have large $M_r(\infty)$ / M_s due to the large number of easy axes of magnetization. Since the resulting ratio of M_s^2 / K_1 is small, interaction effects are weak and the ARM curves have extended and steep linear portions, as expected. The T-dependence of H_c is felt to be a problem. Other perpendicular recording media will be discussed later.

(d) CrO_2 Particles.

These are produced by hydrothermal or other methods. Additives include Ir giving particularly interesting properties. The particles are beautifully shaped without surface imperfections or internal cavities, giving coercive forces over 600 oersted.

(e) Metal Particles, Fe.

These may be prepared by reduction of an oxide or other methods. Pyrophoricity is avoided by coating, and long-time research and development was necessary to produce these particles whose large saturation magnetization leads to coercive forces up to about 2000 oersted.

(f) Barium Ferrite, BaO.6 Fe_2O_3.

Hexagonal platelets (diameter ~ 0.08 μ) were produced by rapid quenching and recrystallization, giving media with perpendicular anisotropy to their plane and thus potential for perpendicular recording. The relevant coercive force is typically 1250 oersted and the relevant $M_r(\infty)$ / M_s > 0.9.

(g) Co-Cr Films.

These are produced by sputtering and contain Co-rich columnar crystallites perpendicular to the film plane separated by less magnetic grain boundaries. These potentially useful perpendicular recording media behave like fine particle materials with important interaction effects which govern, for example, the observed domain walls of the film as a whole ("interaction domains"). Coercive forces up to 2000 oersted have been observed as a result of the fine particle nature of these media. Other thin films (Co-P, Co-Ni, ...) are similar [7,8].

MINT research is most rewarding in all respects and should be encouraged as strongly as possible.

(iv) Permanent Magnets. [9]

These are also materials of enormous potential; all aspects of fine particle magnetism are again involved. For example, ARM curves give information on alloy structure and studies of time effects are relevant to thermal stability. Not all permanent magnets are simply fine particle materials, since in some cases exchange-controlled domain-wall motion and pinning is involved. Shape-anisotropy-controlled permanent magnets include Alnico and crystal anisotropy governs the magnetic hardness of barium ferrite, $SmCo_5$, and $Fe_{14}Nd_2B$. The last material is the result of a revolutionary breakthrough which led to the production of this tetragonal compound with enormous crystal anisotropy. It is not clear (1986) if the magnetic hardness relies entirely on fine-particle magnetism.

(v) Living Organisms. [10]

Magnetotactic bacteria were found to contain chains of single domain Fe_3O_4 particles in strong interaction to provide a natural compass. Similar effects occur in pigeons and bees, and perhaps in *homo sapiens*.

Acknowledgments.

I am grateful to J. L. Morán-López and L. M. Falicov for inviting me to this wonderful meeting and to the British Council for financial support.

References.

1. A. R. Corradi and E. P. Wohlfarth, IEEE Trans. Mag. 14, 861 (1978).
2. S. Shtrikman and E. P. Wohlfarth, Phys. Lett. 85A, 467 (1981).
3. A. Lyberatos, E. P. Wohlfarth, and R. W. Chantrell, IEEE Trans.Mag. 21, 1277 (1985).
4. D. Chowdhury and A. Mookerjee, Phys. Repts. 114, 1 (1984).
5. D. J. Dunlop, Phys. Earth Planet. Int. 26, 1 (1981).
6. C. Radhakrishnamurty *et al.*, Earth Planet. Sci. Lett. 7, 389 (1970).
7. G. Bate, Handb. Ferromag. Mat. 2, 381 (1980).
8. E. P. Wohlfarth, phys. stat. sol. (a) 91, 339 (1985).
9. R. A. McCurrie, Handb. Ferromag. Mat. 3, 107 (1982); H. Stäblein, *ibid* 441. Volume 4 of this series will contain further articles on permanent magnets.
10. R. P. Blakemore and R. B. Frankel, Sci. American 245, 42 (1981).

Chemical Bonding and Local Structure Effects in the Moment Formation and Magnetic Susceptibility of Metallic Molecules and Compounds

J. Keller and M. Castro

Facultad de Química, Universidad Nacional Autónoma de México, Ciudad Universitaria, Delegación Coyoacán, 04510 México, D.F., México

Local structure and chemical bonding in intermetallic compounds may induce a highly anisotropic magnetization or magnetic susceptibility. With the use of multiple-scattering methods and a local-density functional for exchange and correlation effects, we analyze the electronic structure and magnetic susceptibility of some systems: the Ni_2 and NbIr molecules, a Ni_9 cluster and some transition-metal and rare-earth compounds. In some cases a permanent local magnetic moment develops, and additional interactions between orbital magnetism and spin susceptibility must be included. This additional term of the Hamiltonian gives rise to a Fermi-liquid-like behavior of the f-electrons in rare earths, and incipient similar effects for the d-electrons. Numerical results for all cases are presented.

1. Introduction.

We discuss two different local approximations to the study of electronic and magnetic properties of materials: a local approximation to the electronic exchange and correlation and to the susceptibilities, and a study of the effect of the local atomic distribution on the properties of a surface, defect or ordered bulk material. This second approach can be specified by saying that we try to separate the effects of local order from those arising from long-range order.

In section 2 we briefly review some aspects of density-functional theory, with special emphasis on those details which allow a better understanding of exchange and correlation effects. In section 3 we describe the electronic structure of some small nickel cluster systems. All these clusters have nickel-nickel interatomic distances similar to the nearest-neighbor separation in nickel metal.

2. Local Approximations in the Application of Density-Functional Theory.

Density-functional theory was introduced by Hohenberg and Kohn [1], and practical procedures for its implementation were developed by Kohn and Sham [2]. A self-contained review is presented in Reference [3]. For our purposes the electronic density is written as a sum of the squares of orthonormal auxiliary functions ϕ_k, with weights n_k, and inserted in the Kohn-Sham equations,

$$\rho = \sum_{k=1}^{N} |\phi_k|^2 n_k \quad , \qquad \hat{H}_{KS}\,\phi_k = \varepsilon_k^{KS}\,\phi_k \quad , \tag{1}$$
$$\hat{H}_{KS} = -(1/2)\,\nabla^2 + V + \Phi + (\delta E_{xc}/\delta\rho) \quad .$$

This procedure permits the self-consistent calculation of the density ρ and of some auxiliary energy eigenvalues related, but not equal, to the ionization potentials μ_i:

$$\mu_i = \varepsilon_i + \Delta\mu_i \quad , \qquad \delta\langle E[\rho] - \mu\{N[\rho] - N\}\rangle = 0 \quad . \tag{2}$$

Several points are worth stressing: (i) there is no restriction on the type of auxiliary function to be used in (1); (ii) the corrections to the Kohn-Sham eigenvalue to obtain the chemical potentials can be large for atomic-like and localized orbitals, which is very often the case in both insulators and metallic materials; (iii) if the parameters n_k are originally set to satisfy some symmetry and statistics (occupation) requirements, the Kohn-Hohenberg variational principle will minimize the total energy for that particular symmetry and occupation, i.e. for well defined constraints and not globally. The parameters n_k are not necessarily restricted to be integers, provided that the total number of electrons is the correct one. We call calculations with fractional n_k the fractional-occupation approach. They correspond to a first approximation to the configuration-interaction (CI) procedure. This approach is clear in the examples presented and is standard procedure in most condensed-matter calculations.

3. Molecular and Cluster Calculations.

As a first example of the ideas and techniques used in the study of metallic systems with a few atoms, we present a calculation for the Mo_2 diatomic molecule [4]. The absolute minimum of the energy is obtained at an interatomic distance of 4.06 Bohrs and a local (s-s bonding) minimum is obtained at 5.4 Bohrs if the $1\delta_u$ state, of localized d-electron character, is left empty and a bonding $2\sigma_g$ state, made out of s-electron orbitals, is occupied. This procedure shows the general trend that very localized states could appear below the maximum occupied (auxiliary) energy eigenvalue and remain empty in order to minimize the total energy. As a consequence of this effect the Mo_2 molecule is diamagnetic; otherwise a paramagnetic state should have been found [5].

A more complicated magnetic interaction is found in the NbIr heteronuclear diatomic molecule, where many energy minima are seen as a function of the interatomic distance (Fig. 1).

At the NbIr equilibrium distance, the magnetic moment of the molecule, $2\mu_B$, is unevenly distributed. It is found as a residual magnetization of the niobium atom;

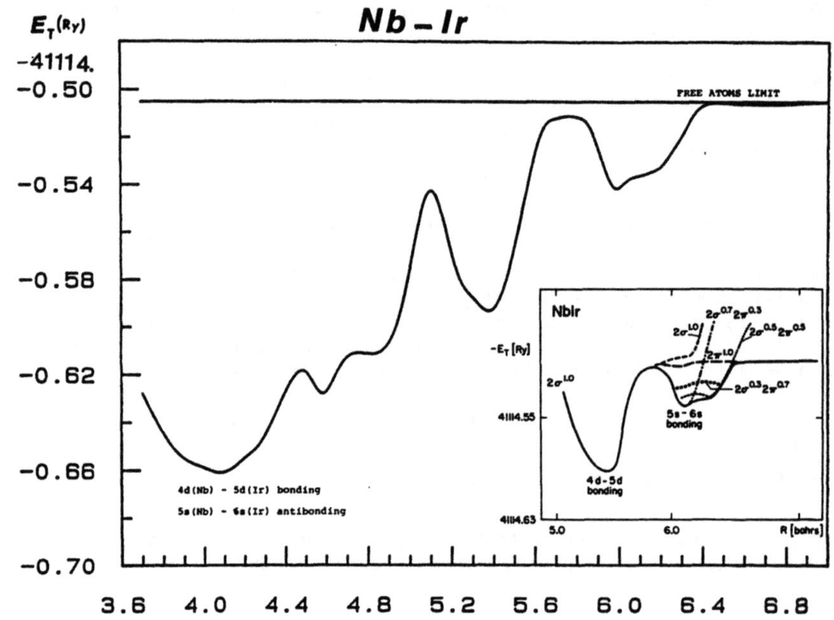

Fig 1. The NbIr molecule total energy as a function of internuclear distance. The various minima correspond to different bonding steps. Fractional occupation of molecular orbitals are used, as shown in the insert.

any magnetization of the iridium atom is antiferromagnetically oriented with respect to that of the niobium atom. The situation corresponds to an infinite-separation limit [6] where the magnetization of Nb ($4\mu_B$) is opposite to that of the iridium d-shell ($-2\mu_B$).

Another example is that of a small cluster of nickel atoms. The original motivation for this study was the possibility of gaining some understanding of the catalytic activity of a small group of metallic atoms deposited on a substrate. For this reason the nickel-nickel interatomic distance was initially set equal to that of bulk nickel metal [7].

The magnetization of the Ni_2 diatomic molecule is straightforwardly given by the occupation of the molecular orbitals and is $1\mu_B$ per atom. The free-atom value is $2\mu_B$. There is nevertheless a new feature in the Ni_2 molecule: the simple approach of filling up orbitals in order of increasing energy eigenvalue (as is traditionally done in the O_2 molecule, for example) would give the correct magnetization, but the splitting of the molecular orbitals caused by the exchange interaction is larger than the separation of the levels, and a final paramagnetic state would have been obtained even without level degeneracy. The exchange splitting for the d-electrons is twice as large as that for the s- and p-electrons in the same molecule. If an H_2 molecule is brought to the proximity of the Ni_2 group with the H_2 and Ni_2 molecular axes parallel to each other, the bonding between the hydrogen atoms is weakened and new bonding appears between the hydrogens and the nearest nickel atoms. This is the onset of an activation mechanism. Other symmetries for this process would be forbidden in a first approximation because of the molecular symmetry conservation rules [8].

The next example is a cluster of nine nickel atoms in the geometry of a half face-centered-cubic cell: an atom with its four nearest neighbors in the (100) plane and its four nearest neighbors in the (100) plane immediately below. The calculation is a cellular, multiple scattering, molecular-like, all electron calculation by means of the $X_{\alpha\beta}$ statistical approximation. Previous calculations for six- and fourteen-atom clusters have been reported [9]. The upper central atom, with a coordination number of 8, acquires a magnetization of (-0.18) spa (spa = spins per atom). The other four atoms in the upper plane, with a coordination of 3, have a magnetization of (+0.563) spa each. The five atoms in the plane below, with a coordination number of 5, have a magnetization of (+0.415) spa each.

In the small-cluster calculations the following trend can be detected: (i) the free Ni atom has a magnetization of 2 spa; (ii) the atoms in the Ni_2 molecule have a magnetization of 1 spa, corresponding to two unpaired molecular orbitals; (iii) the Ni_9 cluster has four electrons in unpaired molecular orbitals, with an average magnetization of 0.44 spa, but with an uneven distribution of magnetization (the magnetization decreases with the number of nearest neighbors). Further study of the Ni_9 cluster shows that the central atom is unstable against an antiferromagnetic transition, with an energy change of $E = (-0.018)$ Ry if the spin of this central atom is reversed. This coupling is stronger that that found in bulk fcc nickel. The Ni_9 cluster is unstable in the studied geometry; it can only exist if trapped in a matrix or stabilized on a surface of, for example, a transition-metal transition-metal or a rare-earth transition-metal alloy. In any case chemical bonding effects dominate over intraatomic exchange effects, but the intraatomic exchange is strong enough to impose an overall magnetization on the cluster. In bulk nickel [9] our calculations yield a magnetization very close to that found in band-structure calculations or experiments: a value of 0.58 spa. Defects, surfaces and small clusters of nickel must therefore show a larger magnetization per atom, with interesting ferro- and antiferromagnetic couplings.

4. Anisotropy of the Magnetic Susceptibility and of the Magnetization.

A transition-metal or a rare-earth atom in an intermetallic dilute alloy may show a local moment even if the host metal is diamagnetic. A common procedure to investigate the formation of local moments in a crystalline, ordered, pure metal or alloy consists of the calculation of two parameters: (i) the density of states at the Fermi level, and (ii) the average exchange enhancement parameter. If the product of these two quantities is larger than unity, then the system is magnetically unstable and a new structure (ferro-, antiferro- or ferrimagnetic) is to be searched for. In dilute alloys, on the other hand, the average density of states or the average exchange enhancement can be very small but the isolated impurity can acquire a magnetic moment neverthelesss. It is therefore evident that a local approach should be used. In addition, current research on spin-fluctuating systems shows that, in many cases, the magnetically unstable region is either in the interstitial or the bonding regions of the crystal. We have generated a formalism to calculate the spin susceptibility in a local approximation [10]

$$\chi_s(\vec{r}) = 2\mu_B^2\, \gamma(\vec{r}) \;/\; [1-\gamma(\vec{r})I(\vec{r})] \quad ,$$

where $\gamma(\vec{r})$ is the local density of electrons with energy equal to the Fermi level, and $I(\vec{r})$ is a local exchange field computed from the second derivative of the exchange and correlation potential with respect to the local magnetization. Our approach is, in all other respects, parallel to that used by Janak and Gunnarsson [11] to study the uniform spin susceptibility.

This approach can be extended to study [12] the total susceptibility, $\chi = \chi_s + \chi_C + \chi_D + \chi_L$, which is the sum of the average value $\chi_s = \langle\chi_s(\vec{r})\rangle$ of the spin enhanced susceptibility, the core-electron diamagnetic susceptibility χ_C, the conduction-electron diamagnetic susceptibility χ_D, and the paramagnetic susceptibility χ_L of the localized magnetic moments. As discussed below, the properties of transition, rare-earth, and actinide metals can depend strongly on the local spin enhanced susceptibility $\chi_s(\vec{r})$.

As examples we have chosen, among others, a spin paramagnet (metallic Nb), a ferromagnet (body-centered-cubic Fe), and a spin fluctuator ($TiBe_2$). Our results are internally consistent, in the sense that paramagnetic metals like Nb present a diffuse, small-value, spin enhanced susceptibility; the average value agrees with experiment. Ferromagnetic iron shows a localized instability which gives rise to a localized magnetic moment in each iron atom; once the material is allowed to acquire a local magnetic moment in a selfconsistent way, there is no longer an instability in the spin susceptibility. In the case of the spin fluctuator $TiBe_2$ we have found [13] that $\chi_s(\vec{r})$ diverges in the region of the bonds between the atoms, but there is frustration between the tendency to acquire a magnetic moment and the wavelengths of the electrons at the Fermi level, which tend to induce a RKKY antiferromagnetic coupling between nearest-neighbor atoms. As a result the tendency to form local moments is suppressed unless an extra factor (an external field or a Jahn-Teller distortion, for example) stabilizes some particular electronic structure.

In this context a divergent itinerant spin susceptibility in a material (rare earth or actinide) that can acquire a permanent magnetic moment may produce a new interaction [10]: a coupling between a localized electron orbit and an itinerant electron spin.

Magnetic coupling between conduction and f electrons modify the kinetic energy of the f electrons. In this respect the conduction electrons must be conceived as a source of magnetic field, and that field incorporated into Schrödinger's equation in the form of a vector potential (f-electron spins should also contribute)

$$[1/2m] \sum_i [p_i - (e/c)\, A_i]^2 \;\approx\; [1/2m^*] \sum_i p_i^2 \text{ , where } m^* - m \approx 2[\Delta E_{so}/E_k]^{1/2}.$$

The effective magnetic field H can be approximated by an effective-mass correction arising from the spin-orbit splitting because $\Delta E_{so} \approx gH$. It should also be noted that the effective field could be larger because the spin-orbit interaction in the lowest state aligns the spin of the f-electrons antiparallel to the orbital magnetic moment, and the Curie field, because of the exchange forces between the f-electron spins and the conduction-electron spins, adds to the dipole field of the orbital moment.

To illustrate the effect of this effective mass correction we have computed the change in the f-band width, Γ_f, in γ-Ce as a function of the effective mass m*. This is shown in Table 1, together with the d-band width, Γ_d, and the relative positions of the centers of the bands, ε_f and ε_d, relative to the bottom of the conduction band.

Table 1. Effect of effective mass correction on band widths and centers

m/m*	$\varepsilon_f - V_{MTO}$ Ry	Γ_f Ry	$\varepsilon_d - V_{MTO}$ Ry	Γ_d Ry
0.98	1.048	0.099	0.90	0.48
0.99	0.845	0.045	0.84	0.45
1.00	0.619	0.016	0.77	0.42
1.01	0.362	0.003	0.71	0.38
1.02	0.072	0.001	0.64	0.35

We found that the f wavefunctions contract as m* increases and acquire an atomic-like character. This is an important property, because it indicates the onset of a cooperative effect: the interaction between the orbital magnetic moment of the f electron and the spin of the conduction-band electrons generate an enhanced effective mass for the f electrons, which in turn, by contraction of the wavefunction, enhance the atomic-like character of the f-band. This is a highly non-linear process which tends to produce sudden changes from f electrons with an itinerant, band behavior to an atomic-like, heavy-fermion behavior. This many-body effect is reponsible for further scattering of the conduction electrons from the rare-earth or actinide ions, which in turn may produce unusual magnetoresistance, Kondo-like behavior, and the possibility of a very special S=1 electron-electron superconducting pairing, which may be characteristic of these materials [14,15].

References.

1. P. Hohenberg and W. Kohn, Phys. Rev. 136, 864 (1964).
2. W. Kohn and L. J. Sham, Phys. Rev 140, A1133 (1965).
3. J. Keller and C. Amador, Density Functional Theory -Lecture Notes in Physics 187, ed. by J. Keller and J. L. Gázques (Springer-Verlag, Berlin-Heidelberg-New York-Tokyo, 1983), p. 269.
4. M. Castro, J. Keller and P. Mareca, Intern. J. of Quant. Chem.: Quantum Chem. Symp. 15, 429 (1981); J. Keller, Rev. Soc. Quím., Mex. 26 (1982).
5. M. M. Goodgame and W. A. Goddard III, Phys. Rev. Lett. 48, 135 (1982); B. Delley, A. J. Freeman and D. E. Ellis, Phys. Rev. Lett. 50, 488 (1983).
6. M. Castro, Thesis, Faculty of Chemistry, Universidad Nacional Autónoma de México, 1984 (unpublished).
7. J. Keller, M. Castro and A. L. de Paoli, J. Appl. Phys. 53, 8850 (1982).
8. R. Hoffmann and R. B. Woodward, The Conservation of Orbital Symmetry (Academic Press New York, 1970).
9. J. Keller and M. Castro, J. Magn. Magn. Mater. 15-18, 856 (1980).
10. J. Keller, C. Amador and C. de Teresa, Physica 130B, 37 (1985).
11. J. F. Janak, Phys. Rev B 16, 255 (1977), and references therein.
12. J. Keller, Hyperfine Interactions 6, 15 (1979)
13. C. de Teresa, J. Keller and J. Schoenes, Solid State Comm. 56, 871 (1985).
14. P. W. Anderson Phys. Rev. B 30, 4000 (1984).
15. J. P. Rodríguez, Phys. Rev. Lett. 55, 250 (1985).

Part VI

Quasi-One-Dimensional Systems

Quasi-One-Dimensional Density-Wave Systems in High Magnetic Fields

C.A. Balseiro and L.M. Falicov

Department of Physics, University of California, Berkeley, CA 94720, USA

The effect of high magnetic fields on the electronic structure of quasi-one-dimensional systems with charge- or spin-density waves is examined. Because of the anisotropy of the system, small electron and hole pockets created by the density wave can be destroyed by the presence of a magnetic field. This phenomenon is analyzed for various orientations of the field. The energy spectrum and the density of states in the presence of an external field are calculated for a simple quasi-one-dimensional model band structure. The effect of the magnetic field on the density-wave critical temperature is also calculated.

1. Introduction

The problem of the behavior of charge- and spin-density waves (CDW and SDW) in high magnetic fields is currently under close scrutiny [1-6]. Although these two kinds of broken symmetry states are very different -- CDWs are caused by the electron-phonon interaction and produce a charge redistribution throughout the crystal whereas SDWs are induced by the repulsive electron-electron interaction and produce a spatiallly varying spin polarization -- the effect of a high magnetic field on the electronic structure of CDWs and SDWs is, to a great extent, similar.

The effect of magnetic fields on the SDW systems $(TMTSF)_2X$, where X stands for PF_6 or ClO_4 has been studied experimentally [1,7,8] and theoretically [2,3]. These compounds are regarded as examples of quasi-one-dimensional (Q1D) systems. At low temperatures they exhibit a stable SDW state. Under applied pressure, however, the SDW is destroyed, a consequence of changes in the electronic structure and in the Fermi surface. It has been reported that the application of a magnetic field perpendicular to the axis of high conductivity restores the SDW in the high pressure regime.

GOR'KOV and LEBED [2] gave the first interpretation of this remarkable effect. They showed that for a simple anisotropic metal with two open Fermi surface sheets (corrugated planes), the tendency to SDW formation is enhanced by a magnetic field applied in a direction parallel to the sheets.

HÉRITIER et al. [3,9] generalized the theory by including a magnetic-field dependence of the SDW $\vec{q}$-vector. They found a cascade of first-order transitions that could be observed as the magnetic field is increased [8].

These theoretical contributions focused only on the instabilities of the normal phase, i.e. they studied the divergences of the generalized susceptibilities in the normal phase. The susceptibility $\chi_0(\vec{q},\vec{H},T)$ for various values of the $\vec{q}$-vector, the magnetic field $\vec{H}$ and the temperature T was evaluated for some simple band structures [9]. Friedel [6] discussed these models in semiclassical terms, which is only valid for weak magnetic fields. He argued that the same effects should be observed in nearly stable CDWs as well.

Recently, an anomalous magnetic-field dependence of the transport properties in the low-temperature CDW phase of $NbSe_3$ has been reported [4]. It is also known that $NbSe_3$, a Q1D system, exhibits two phase transitions [10-12] associated with CDWs at T_1=144 K and T_2=59 K. In the low temperature phase, a magnetic field perpendicular to the high-mobility b-axis produces an anomalous magnetoresistance. The anomaly has been interpreted as a substantial decrease in the number of carriers caused by the magnetic field, i.e. a magnetic-field induced obliteration of parts of the Fermi surface [4,5]. It has been shown that because of the anisotropy of the system, small electron and hole pockets created by the CDW can be destroyed by the effects of the magnetic field, resulting in an effectively better nesting of the Fermi surface.

In what follows we discuss the effect of high magnetic fields on the electronic structure and the critical temperature of Q1D systems with either CDWs or SDWs.

2. Density Waves in a Magnetic Field

The model used here to represent the Q1D system consists of a tetragonal lattice with constants $a_1 \ll a_2 = a_3$ in which an electronic single s-like tight-binding band exists. The dispersion relation is given by

$$E_0(\vec{k}) = -2t \cos(k_x a_1) - 2t' [\cos(k_y a) + \cos(k_z a)]$$
$$-4t'' \cos(k_y a) \cos(k_z a) , \qquad (1)$$

where $\vec{k} = (k_x, k_y, k_z)$ and $a = a_2 = a_3$. The parameters t and t' are the electron-transfer matrix elements between nearest neighbors in the x-direction and in the (y,z) plane respectively; t" is the electron-transfer matrix element between second-nearest neighbors in the (y,z) plane. Because of the Q1D character of the system $t \gg t', t''$.

Except for energies close to either the bottom or the top of the band, the surfaces of constant energy in $\vec{k}$-space are given by two open, corrugated sheets. The system shows a tendency to form density waves (DW) with a wavevector $\vec{q}$ which connects (nests) the two open sheets of Fermi surface. The stable DW is either a CDW or a SDW, depending on whether the electron-phonon or the repulsive electron-electron interaction dominates [13].

If the electron density is taken to be one electron per site, the two pieces of Fermi surface are centered at $k_x \cong \pm (\pi/2a_1)$. For $t'' \ll t'$ the DW instability occurs for $\vec{q} = (\pi/a_1, \pi/a, \pi/a)$. The DW is characterized by a gap order parameter τ. Because of the extra periodicity induced in the system by the DW, the one-electron state $|\vec{k}\rangle$ is mixed with the state $|\vec{k}+\vec{q}\rangle$; the matrix element of this mixing is proportional to τ, a quantity assumed to be independent of $\vec{k}$.

The total Hamiltonian in second quantization takes the form

$$H = H_0 + H_1(\tau) \quad , \quad \text{where} \qquad (2)$$

$$H_0 = \sum_{\vec{k}\sigma} E_0(\vec{k})\, c^{\dagger}_{\vec{k}\sigma} c_{\vec{k}\sigma} \, , \qquad (3)$$

$$H_1 = \sum_{\vec{k}\sigma} \tau_\sigma\, c^{\dagger}_{\vec{k}\sigma} c_{(\vec{k}+\vec{q})\sigma} \, , \qquad (4)$$

the operator $c^{\dagger}_{\vec{k}\sigma}$ creates an electron with crystal momentum $\vec{k}$ and spin σ, and

$$\tau_\uparrow = \tau_\downarrow = \tau \qquad \text{for CDWs,}$$

$\tau_\uparrow = -\tau_\downarrow = \tau$ for SDWs.

The DW quasi-particle spectrum of the system is given by

$$E_\pm(\vec{k}) = -4t'' \cos(k_y a)\cos(k_z a)$$
$$\pm[\{2t\cos(k_x a_1) + 2t'[\cos(k_y a) + \cos(k_z a)]\}^2 + \tau^2]^{1/2}. \quad (5)$$

For t"=0 this energy spectrum has a gap 2τ at the center of the band, i.e. at the Fermi level. The DW phase is therefore a semiconductor. This fact is a consequence of the perfect nesting of the normal, $\tau = 0$ Fermi surface.

For $t'' \neq 0$, there is no longer perfect nesting. As before, the DW induces a <u>direct gap</u> of value 2τ. But the spectrum has now a smaller <u>indirect gap</u> of value $2\Delta_g$ given by

$$\Delta_g = \begin{cases} \tau - 4t'' , & \text{if } \tau/t'' > 4 \; ; \\ 0 , & \text{otherwise.} \end{cases} \quad (6)$$

In this case, depending on the value of τ/t'', the system in the presence of a DW is either a semiconductor or a semimetal.

When a magnetic field is applied new complications arise. Whereas the DW causes gaps in an otherwise continuous spectrum, the magnetic field quantizes the electronic motion in the two directions perpendicular to it and tends to form either discrete levels and narrow bands if the cyclotron electronic orbits are closed, or complicated continua if they are open [14-18]. The competition between these various effects are the cause of the strange, complex behavior under study here.

For low magnetic fields, a simple semiclassical description of the motion of the electrons can be used. This scheme, according to Onsager's quantization rules [14], yields only allowed closed electron and hole orbits which enclose an integral number of magnetic field flux quanta. Open orbits, which arise from the combined effect of magnetic field and periodic potential, form a continuum and are not quantized in the traditional sense. This approach completely neglects several secondary effects: band broadening, tunneling between bands, and the complicated structure of the continuum. These effects are of considerable importance in the case under study. Were it not for the presence of the gap parameter τ, the spectrum of the model would yield only open orbits, i.e. a continuum in Onsager's scheme; the gap parameter τ, on the other hand may be of the same order of the magnitude of the cyclotron energy $\hbar\omega_c$ of the closed electron or hole orbits (induced by τ) for reasonably attainable values of the magnetic field.

An acceptable scheme to describe the DW and the effects of the magnetic field on equal footing can be obtained from the well known effective-mass theory of LUTTINGER and KOHN [19]. If paramagnetic (spin) effects are neglected, the Hamiltonian of the system in first quantization is given by

$$H = H_0(\vec{\kappa}) + H_1(\tau) , \quad (7)$$

here the first term is obtained from (1)

$$H_0(\vec{\kappa}) = E_0(\vec{k} = \vec{\kappa}) , \quad (8)$$

with the differential operator $\vec{\kappa}$ defined by

$$\vec{\kappa} = -i\vec{\nabla} + (|e|/\hbar c)\vec{A}(\vec{r}) , \quad (9)$$

and the second term describes the perturbations induced by the DW. For a uniform magnetic field $\vec{H} = (0,H_y,H_z)$ perpendicular to the x-direction, the vector potential takes the form

$$\vec{A}(\vec{r}) = (0,\ -H_z x,\ H_y x)\ . \tag{10}$$

In this gauge, the total Hamiltonian in second quantization can be written

$$H = \sum_{nk_\rho\sigma} \varepsilon(n,k_\rho)\, c^\dagger_{nk_\rho\sigma} c_{nk_\rho\sigma} - t \sum_{nk_\rho\sigma} \sum_{\delta=\pm 1} c^\dagger_{nk_\rho\sigma}\, c_{(n+\delta)k_\rho\sigma}$$
$$+ \sum_{nk_\rho\sigma} \tau_\sigma \cos(nq_x a_1)\, c^\dagger_{nk_\rho\sigma}\, c_{n(k_\rho+q_\rho)\sigma}\ . \tag{11}$$

Here k_ρ (q_ρ) is a two dimensional vector with components k_y and k_z (q_y, q_z). The index n refers to the atomic planes perpendicular to the x-direction and the operator $c^\dagger_{nk_\rho\sigma}$ creates an electron in the state $|nk_\rho\sigma\rangle$ of energy $\varepsilon(nk_\rho)$ given by

$$\varepsilon(nk_\rho) = -\,2t'\,[\cos(k_y a+\Phi_z n) + \cos(k_z a+\Phi_y n)]$$
$$-\ 4t''\cos(k_y a+\Phi_z n)\cos(k_z a+\Phi_y n)\ , \quad \text{where} \tag{12}$$

$$\Phi_I = (|e|/\hbar c)\, H_I\, a_1\, a\ , \qquad (\ I = y,z\ )\ . \tag{13}$$

Results for the energy spectrum for magnetic fields along the [001] and [011] directions are shown in the next section.

3. Energy Spectrum

From the Hamiltonian obtained in Sec. 2, it is clear that the magnetic field induces a new periodicity in the system. The problem of Bloch electrons in a magnetic field has been studied intensively and extensively [14-18]. The physical properties of the system do not depend, except on an extremely fine scale, on the commensurability of the magnetic-field-induced and the lattice periodicities. Within this framework it is possible then, without loss of generality, to restrict oneself to the case of "rational" fields, in particular to values of Φ_I equal to $2\pi/N$, where N is a large integer.

The energy spectrum for each value of k_ρ can be obtained by evaluating the eigenvalues of a transfer matrix [5]. The total energy spectrum is a superposition of the spectra obtained for all values of k_ρ in the two-dimensional Brillouin zone.

For a magnetic field along [001] the two-dimensional (2D) energy spectra for E>0 and various values of k_z are shown in Fig. 1. The spectra for E<0 are very nearly the same. The numerically exact method employed in the calculation is slowly convergent and no longer practical for small values of the magnetic field (N>2000). Interpolation between the lowest calculated values and H=0 is easily accomplished, and is shown as an example in Fig. 1(a). Results can be summarized as follows:

(i) For low magnetic fields the semiclassical spectra are reproduced.

(ii) For intermediate and high fields there is a substantial broadening of all Landau levels.

(iii) For high magnetic fields electron and hole levels are very broad, and the lowest (highest) boundary of the broadened n = 0 electron (hole) level is at an energy very close to τ ($-\tau$). As a consequence, the total three-dimensional spectra for high magnetic fields have a gap for energies $|E|<\tau$ and a continuum of states for $|E|>\tau$. All k_z sections contribute no states for $|E|<\tau$ and continuum states for $|E|>\tau$, just as in the perfect nesting case.

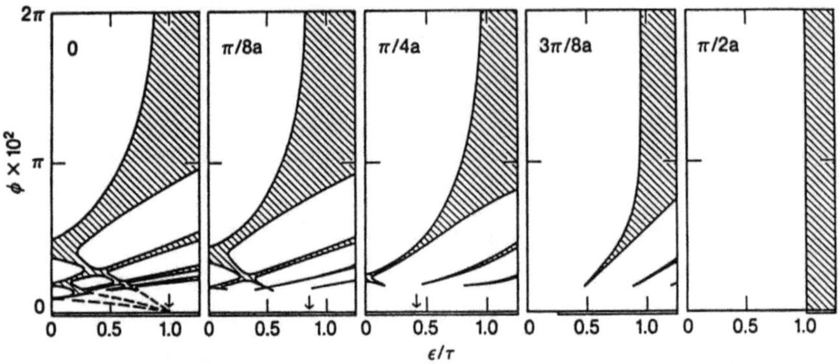

Figure 1.- Energy spectra for magnetic fields parallel to [001], E > 0 (the conduction band) and various values of k_z. The abscissae are energies and the ordinates the magnetic field strength. Extrapolation to $H \to 0$ of the first few hole levels (dashed lines) is shown for $k_z = 0$. The broadened Landau levels can be recognized in all sections.

(iv) In the energy region where all orbits are open the energy spectrum shows broad levels (bands) separated by small gaps. The semiclassical gapless continuum is only recovered for very high energies.

The very large broadening of the Landau levels, in particular the n=0 levels, and the existence of gaps in the semiclassical continuum can be understood only in terms of a magnetic-field induced mixing of electronic bands, i.e. a hybridization of the valence and conduction bands. This mixing is important if the magnetic cyclotron energy $\hbar\omega_c$ is of the same order of magnitude of the gap parameter τ. In fact, the proper scale to measure the magnetic field strength is not Φ_I (which measures the ultra fine structure) but the cyclotron energy

$$\hbar\omega_c = \hbar\ |e|\ H\ /\ m^*\ c \tag{14}$$

with an effective mass m* corresponding to the bottom of the DW-state conduction band

$$m^* = (\hbar^2\ /\ 4\ a_1\ a_2)\ [\ \tau\ /\ t''\ (t^2 - 4t'^2)]^{1/2}\ . \tag{15}$$

It can be seen that if the condition $t \gg t' \gg t''$ is maintained, the energy spectrum is only sensitive to the ratios $\hbar\omega_c/t''$ and τ/t''.

For a high magnetic field the spectrum has a large gap at the center of the band, i.e. the Fermi level. This central gap $2\Delta_g$ is of the order of 2τ, even for values of $\tau/t'' \leq 4$. As the magnetic field is reduced, the size of the central gap decreases and vanishes for a given value of the field which depends on τ. For magnetic fields smaller than this value there is a band of allowed continuum states at the center of the spectrum and the system is semimetallic.

Figure 2 is a diagram of constant energy-gap parameter Δ_g/τ in the ($\hbar\omega_c/t''$, τ/t'') plane. For $\tau/t'' < 4$ there is a two-dimensional region of the plane corresponding to a semimetallic regime, Δ_g=0. The high-magnetic-field region is semiconducting, with the energy-gap parameter increasing monotonically with either variable.

Because of the intricate fine structure of the spectrum, characteristic of Bloch electrons in a magnetic field, a very small gap could appear at the Fermi level in the semimetallic regime for some particular values of the magnetic field. These are in general very small gaps of a different nature

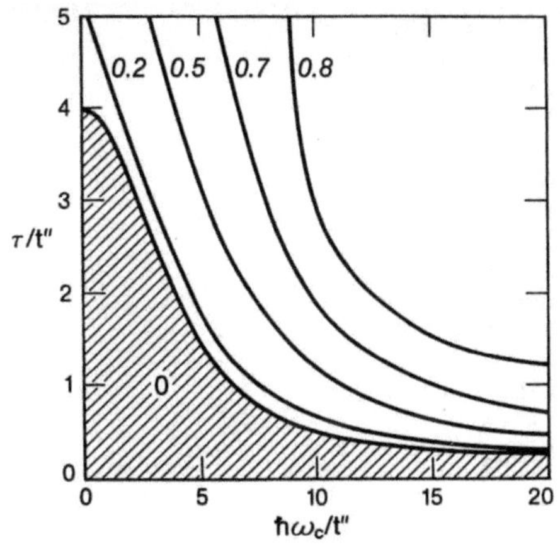

Figure 2.- The values of Δ_g/τ as a function of magnetic field $\hbar\omega_c/t''$ and τ/t'' for $\vec{H}$ parallel to [001], $t'/t = 0.2$ and $t'' \ll t'$. The shaded region corresponds to $\Delta_g = 0$ and band overlap, the semimetallic case. The gap increases rapidly to its maximum value $\Delta_g = \tau$. Note that for all parameter values a semiconducting state is obtained if the field strength is large enough.

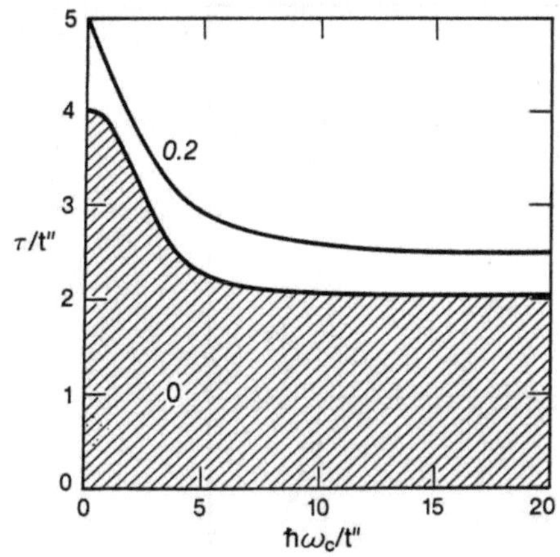

Figure 3.- The values of Δ_g/τ as a function of magnetic field $\hbar\omega_c/t''$ and τ/t'' for $\vec{H}$ parallel to [011], $t'/t = 0.2$ and $t'' \ll t'$. The shaded region corresponds to $\Delta_g = 0$ and band overlap, the semimetallic case. The gap increases rapidly to its maximum value $\Delta_g = \tau$. Note that only for $\tau/t'' > 2$ a semiconducting state is obtained if the field strength is large enough. For $\tau/t'' < 2$ the system is always semimetallic, regardless of magnetic-field strength.

that are completely obliterated by small inhomogeneities in the field and that do not change the physical properties of the system.

A magnetic field along [011] produces similar but less pronounced effects. Figure 3 is a diagram of constant energy gap for this case. Any reasonable high magnetic field produces now a semiconducting state only for $\tau/t'' > 2$.

Energy spectra for magnetic fields along the high-conductivity [100] x-axis were evaluated in the same fashion. For the field in this direction no gaps were obtained if $\tau/t'' < 4$.

It is instructive to study the behavior of the density of states as the magnetic field increases. In what follows numerical results for the density of states in the region of interest, i.e. the center of the band, are presented.

The total density of states per spin is given by

$$D_\sigma(E) = (-1/\pi) \lim_{\eta\to 0} \sum_{n\,k_\rho} \mathrm{Im}\, G_{nk_\rho\sigma}(E+i\eta) \ , \tag{16}$$

where Im indicates the imaginary part and $G_{nk_\rho\sigma}$ is a Green's function, given by

$$G_{nk_\rho\sigma}(E+i\eta) = \langle\langle c_{nk_\rho\sigma}, c^\dagger_{nk_\rho\sigma} \rangle\rangle_{E+i\eta} \ . \tag{17}$$

A continuous fraction method [5] is used to evaluate G. Densities of states at the center of the band for three different values of the magnetic field along [001] are shown in Fig. 4. Figure 4(a) corresponds to a value of the field in the semimetallic regime; a complex structure characteristic of Bloch

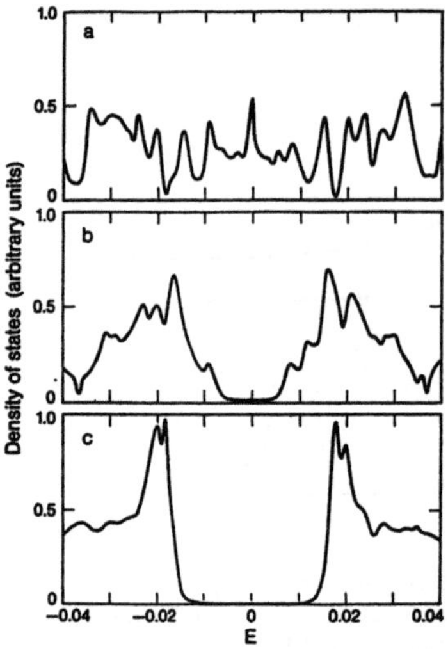

Figure 4.- Total (3D) densities of states near the Fermi level for magnetic fields parallel to [001] and various values of the field strength. (a) Semimetallic regime $\hbar\omega_c/\tau = 2.03$ ($N = 400$); (b) semiconducting regime $\hbar\omega_c/\tau = 4.07$ ($N = 200$) with $\Delta_g/\tau=0.38$; (c) semiconducting regime $\hbar\omega_c/\tau = 8.14$ ($N = 100$) with $\Delta_g/\tau=0.85$.

electrons in a magnetic field is readily apparent. Higher values of the magnetic field are presented in Figs. 4(b) and (c), where Δ_g/τ attains the values 0.38 and 0.85, respectively. The small tails in $D_\sigma(E)$ which can be observed near the gap edges are a consequence of the finite imaginary part η of the energy used in the numerical evaluations. Except for the fine-structure details, the densities of states at high magnetic fields resemble that of the system with perfect Fermi surface nesting, i.e. no k_z-section contributes states to the $|E|<\tau$ region of the spectrum. Contrary to the predictions of the semiclassical model there is a real gap in the spectrum.

4. Critical Temperature

The order parameter τ must be calculated selfconsistently, and is in general a function of the direction and the strength of the magnetic field. The equilibrium value of τ is given by

$$\tau = \lambda \sum_{n\, k_\rho} \cos (nq_x a_1) \quad \langle c^\dagger_{nk_\rho\sigma}\, c_{n(k_\rho+q_\rho)\sigma}\rangle \ , \tag{18}$$

where λ is the interaction parameter and $\langle ... \rangle$ indicates thermodynamic averages. A selfconsistent calculation with the present model is a tedious and numerically expensive proposition which has not been carried out. Two extreme situations can be however readily analyzed. At T=0 the selfconsistent solution of (3.9) can be obtained by minimizing the total energy

$$E_T = \sum_\sigma \int_{-\infty}^{\varepsilon_F} D_\sigma(E)\, E\, dE \quad - \tau^2 / 2\lambda \ , \tag{19}$$

where ε_F is the Fermi level. Since the density of states $D_\sigma(E)$ for high magnetic fields along [001] resembles that characteristic of the system with perfect Fermi surface nesting, the order parameter τ for this limiting case must be numerically close to that obtained in the perfect-nesting case.

The critical temperature and its magnetic field dependence can also be calculated in a simple way. A linearization of the dispersion relation (1) for k_x in the vicinity of $\pm(\pi/2a_1)$ and the procedure decribed in Ref. 2 yields for the magnetic field in the [001] direction the following equation for the critical temperature T_c

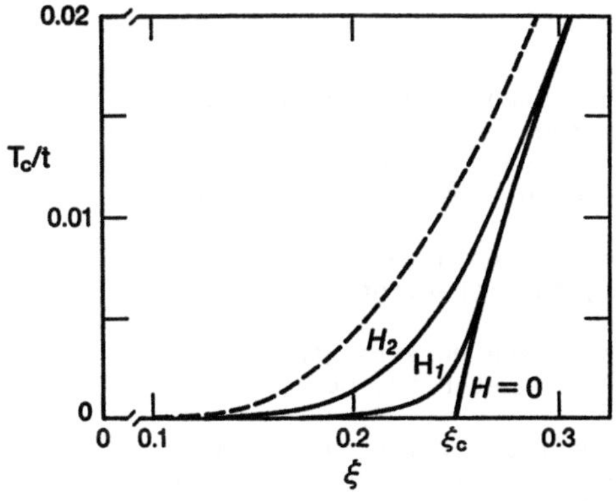

Figure 5.- The critical temperature T_c as a function of interaction strength ξ for (a) the perfect nesting case $t''=0$ and $H=0$ (dashed line); (b) a semimetallic case $t''=0.01\ t$ and $H=0$ (observe the critical value ξ_c); the case shown in (b) for magnetic fields parallel to [001], strength such that $N=400$ (c) and $N=200$ (d). It should be noted that T_c changes appreciably with magnetic-field strength only for ξ in the vicinity of ξ_c. For the lattice parameters of $NbSe_3$, $N=400$ correspond to $H \approx 250$ kG.

$$1 = \frac{\lambda}{4\pi t a_1 a^2} \int_{\pi T_c/t}^{\infty} \frac{dy}{\sinh y} J_0{}^2[(4t''/t\Phi)\sin(yt\Phi/2\pi T_c)] \quad , \qquad (20)$$

where $\Phi = \Phi_z$ of (13) and J_0 is the ordinary Bessel function.

In Fig. 5 the critical temperature T_c as a function of the dimensionless coupling constant $\xi = \lambda/4\pi t a_1 a^2$ is shown. Because of the lack of perfect nesting at H=0 the DW is stable only for $\xi > \xi_c$, a minimum critical value of the interaction. A magnetic field however can induce a DW for $\xi < \xi_c$. This magnetic-field-induced DW was predicted by Gor'kov and Lebed [2]. For $\xi > \xi_c$ the presence of a magnetic field raises the value of T_c. However, if the critical temperature for zero magnetic field is not too low, T_c does not change appreciably with magnetic field. This effect is shown in Fig. 6, where T_c is plotted against magnetic field strength for fields along [001] and for various values of ξ. The critical temperature increases monotonically with the field.

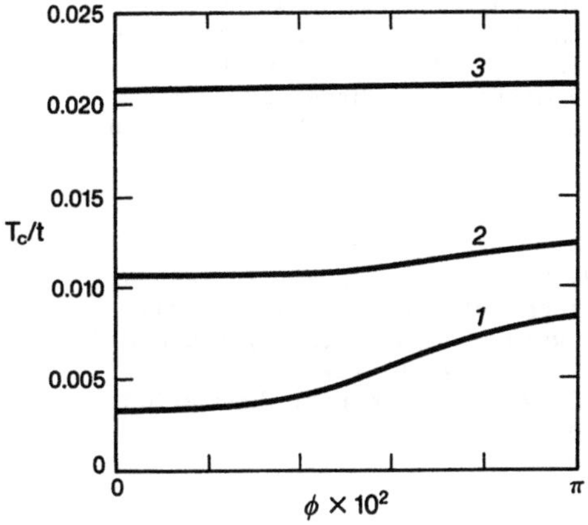

Figure 6.- The DW transition temperature T_c as a function of magnetic field strength for fields parallel to [001] and various values of the interaction parameter ξ. Notice that T_c increases monotonically with field strength, but that the increase is sizeable only for ξ in the vicinity of ξ_c.

For magnetic fields along [011] the 2D band structure becomes equivalent to that used in Ref. 2. The increase in T_c for this direction of the field shows oscillatory behavior. Here again the effect of the magnetic field is important only for values of ξ close to the critical value ξ_c, i.e. for anisotropic DW systems which are either marginally stable or marginally unstable.

5. Summary and Discussion

The effect of a magnetic field on the electronic structure of a Q1D system with a CDW or a SDW has been examined. The system is characterized by a well defined $\vec{q}$-vector and by a lack of perfect nesting which result, at zero field, in a semimetallic system. The results can be summarized as follows:

(a) A magnetic field perpendicular to the axis of high conductivity may induce gaps in the spectrum at the Fermi level, thus transforming a semimetallic structure into a semiconducting one.

(b) The value of the gap Δ_g is a function of the DW strength τ, the cyclotron energy $\hbar\omega_c$ and the "lack of nesting" parameter t", but insensitive to other band-structure parameters.

(c) Although at low magnetic fields the spectrum of the system is well described by the semiclassical quantization scheme, this scheme breaks down for fields such that the cyclotron energy is of the order of, or greater than the DW mixing parameter τ; in the latter regime band mixing becomes paramount.

(d) The complete spectrum of the system as a function of the various parameters shows two main effects: (d1) in the semiclassical (closed orbit) discrete regime the Landau levels spread into wider and wider bands as the field strength increases; and (d2) in the semiclassical (open orbit) continuum energy gaps appear at high fields.

(e) The semimetal-to-semiconductor transition does not take place for magnetic fields parallel to the axis of high conductivity, the x-axis in the present model.

(f) There is also considerable anisotropy for fields perpendicular to the high-conductivity axis [the (y,z) plane]. For fields along the <001> axes semimetal-to-semiconductor transitions are obtained for high enough field strength regardless of the value of the ratio τ/t". For fields along <011> axes, the semiconducting gap does not open up at any field strength if the ratio τ/t" is smaller than a given value, 2 in this model.

(g) The DW transition temperature T_c is enhanced by the magnetic field. The enhancement, however, is only appreciable if the situation is such that the interaction strength ξ is just below or just above the critical value ξ_c, i.e. for marginally stable or marginally unstable [2], semimetallic DWs. For well established, stable DWs the enhancement effect is small.

(h) All effects encountered in this tetragonal model can be directly applied to interpret the experiments in the more anisotropic, real systems such as $NbSe_3$ (a CDW system) or the TMTSF salts (SDWs). The complexity of the spectra and electronic effects found here are complicated even more by the complexity of the real band structure in these materials, where several sheets of Fermi surface (of various topologies) contribute to the conduction processs.

Acknowledgments

The authors are deeply indebted to Prof. R. V. Coleman for many illuminating and challenging discussions, and for keeping them informed on a day-to-day basis of the results of his experiments. This work was supported by the NSF through Grant DMR81-06494. One of the authors (CAB) would like to acknowledge the financial support of a J. S. Guggenheim Memorial Foundation Fellowship and of the Centro Atómico Bariloche (Argentina).

References

1. R. Brusetti, P. Garoche and K. Bechgaard, J. Physique Colloq. C 3, 1051 (1983).
2. L. P. Gor'kov and A. G. Lebed, J. Phys. (Paris), Lett. 45, 433 (1984).
3. Héritier, G. Mantambaux and P. Lederer, J. Phys. (Paris), Lett. 45, 943 (1984).

4. R. V. Coleman, G. Eiserman, M. P. Everson, A. Johnson and L. M. Falicov, Phys. Rev. Lett. 55, 863 (1985).
5. C. A. Balseiro and L. M. Falicov, Phys. Rev. Lett. 55, 2336 (1985); Phys. Rev. B (in press).
6. J. Friedel, Phil. Trans. Roy. Soc. London, Ser. A 314, 189 (1985).
7. T. Takahashi, D. Jérome and K. Bechgaard, J. Phys. (Paris), Lett. 43, 565 (1982).
8. F. Pesty, P. Garoche and K. Bechgaard, Phys. Rev. Lett. 55, 2495 (1985).
9. G. Montambaux, M. Héritier and P. Lederer, Phys. Rev. Lett. 55, 2078 (1985).
10. P. Monceau, N. P. Ong, A. M. Portis, A. Meerschaut and J. Rouxel, Phys. Rev. Lett. 37, 602 (1976).
11. N. P. Ong and P. Monceau, Phys. Rev. B 16, 3443 (1977).
12. R. M. Fleming, D. E. Moncton and D. B. McWhan, Phys. Rev. B 18, 5560 (1978).
13. C. A. Balseiro, P. Schlottmann and F. Yndurain, Phys. Rev. B 21, 5267 (1980).
14. L. Onsager, Philos. Mag. 43, 1006 (1952).
15. P. G. Harper, Proc. Roy. Soc. London, Sect. A 68, 874 (1955).
16. J. Zak, Solid State Phys. 27, 1 (1972).
17. W. Y. Hsu and L. M. Falicov, Phys. Rev. B 13, 1595 (1976).
18. D. R. Hofstadter, Phys. Rev. B 14, 2239 (1976).
19. J. M. Luttinger and W. Kohn, Phys. Rev. 97, 869 (1955).

Theory of the Spin-Peierls Transition in Quasi-One-Dimensional Systems

H. Fukuyama[1] *and S. Inagaki*[2]

[1]Institute for Solid State Physics, University of Tokyo, Roppongi, Minato-ku, Tokyo 106, Japan

[2]Physics Laboratory, Faculty of Engineering, Meiji University, Higashi-Mita, Tama-ku, Kawasaki 214, Japan

A phase Hamiltonian approach is shown to be able to take proper account of quantum fluctuations, which are essential in one-dimensional S = 1/2 Heisenberg antiferromagnets. It is applied to describe the spin-Peierls transition, with particular emphasis on the competition with the magnetic state in the presence of weak interchain exchange interaction. Various possible types of the phase transitions induced by the magnetic field are also pointed out.

1. Introduction.

The one-dimensional S = 1/2 Heisenberg antiferromagnet is unique because of its full quantum-mechanical fluctuations. In spite of the fact that there exist exact results for various physical quantities [1], the physical understanding of this model system is not yet complete. For example, the existence of the linear-in-q spin-wave mode [2] is easily accepted if the ground state is assumed to be the antiferromagnetic (Néel) state and only if the lowest-order contributions in spin-wave operators (in Holstein-Primakoff transformation) are considered [3]. However, fluctuations coming from higher orders are divergent [4] and invalidate the initial assumption of the stability of the Néel state. This fact implies that the ground state consists of spin configurations very different from that of Néel state. In fact ANDERSON [5] pointed out that the ground-state energy variationally determined with wave functions consisting of a liquid of singlet pairs, which he called resonating valence bond (RVB), can come closer to the exact result of HULTHÈN [6]. This finding indicates that the non-magnetic singlet components, i.e. states with spin configurations very different from that of the Néel state, occupy an important fraction of the ground state.

Quantum fluctuations with such features present in the ground state of purely one-dimensional Heisenberg systems lead to three-dimensional ordering in actual systems in the presence of possible interchain exchange interactions among spins and coupling to the three-dimensional lattice. The former stabilizes the Néel state, whereas the latter favors the dimerized singlet state, which is called spin-Peierls (SP) state. (In reality the occurrence of the Néel state is far more prevalent than the SP state for reasons mentioned below.) The existence of the SP state in actual systems, however, offers a unique opportunity to understand the nature of singlet components in the one-dimensional Heisenberg model and to explore further the remarkable insight of Anderson.

In this paper we demonstrate that the phase Hamiltonian approach is very suited to determine the existence of large quantum fluctuations in the one-dimensional Heisenberg model. Based on this approach, the competition between the Néel and the SP states in quasi-one-dimensional systems is examined.
We take throughout units of $\hbar = 1$.

2. Basic Experimental Facts of the Spin-Peierls Transition. [7]

Among S = 1/2 Heisenberg antiferromagnets, systems which exhibits the SP transition have a spin susceptibility, χ, with a temperature dependence as shown in Figure 1; T_{SP} is the transition temperature. An important fact is that χ is essentially

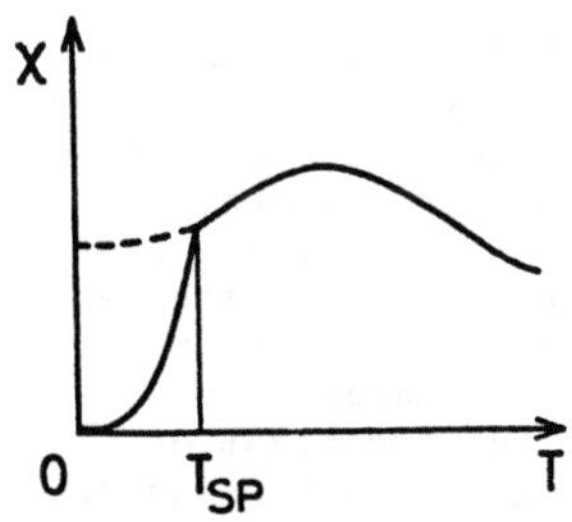

Figure 1. Temperature dependence of χ in spin-Peierls systems.

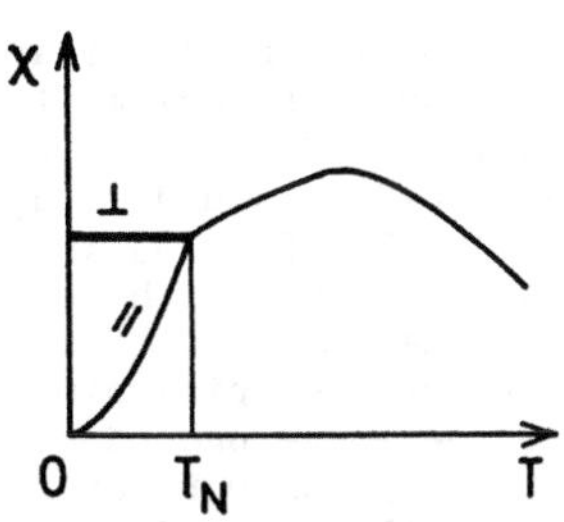

Figure 2. Temperature dependence of χ in quasi-one-dimensional antiferromagnets.

isotropic even below T_{SP} and vanishes as $T \to 0$. In this figure, the broken line is meant to be that of purely one-dimensional spin systems as examined by BONNER and FISHER [8]. The temperature dependence of χ is in contrast to that of systems which undergo an antiferromagnetic transition, as shown in Figure 2; there T_N is the Néel temperature and χ is anisotropic below T_N. Moreover, in the cases of a SP transition there is a dimerization of the lattice, as shown in Figure 3, for $T < T_{SP}$.

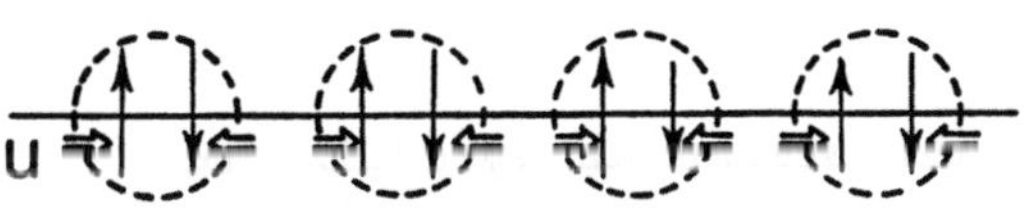

Figure 3. The dimerization, u, in the spin-Peierls state.

3. One-dimensional Heisenberg Antiferromagnets.

The Hamiltonian of the one-dimensional $S = 1/2$ Heisenberg antiferromagnets in an undistorted lattice is given by

$$H_0 = J \sum S_\ell \cdot S_{\ell+1} \,, \tag{1}$$

where the exchange interaction, J, is assumed to be finite only between neighboring sites.

As is well known, one-dimensional spins with $S = 1/2$ can be transformed into spinless fermions by Jordan-Wigner transformation [9],

$$S_\ell^+ = \exp(-i\pi \sum_j^{\ell-1} a_j^\dagger a_j) a_\ell^\dagger = (S_\ell^-)^\dagger, \quad S_\ell^z = a_\ell^\dagger a_\ell - 1/2 \,, \tag{2}$$

which transforms (1) into an interacting fermion system with a Hamiltonian given by

$$H_0 = \sum \varepsilon(k) a_k^\dagger a_k + \frac{1}{N} \sum v(q)\, a^\dagger_{k+q} a^\dagger_{k'-q} a_{k'} a_k \,, \tag{3}$$

$$\varepsilon(k) = -J \cos ka \,, \tag{4}$$

$$v(q) = J \cos qa \,. \tag{5}$$

Note that both the kinetic energy, $\varepsilon(k)$, and the interaction energy $v(q)$ in (3) are characterized by J, which is the only parameter in (1). In the absence of the magnetic field the present fermion band should be half-filled since $\langle S_\ell^z \rangle = 0$.

The interaction term in (3) is conventionally treated in the Hartree-Fock approximation [7]. However, it is now well known that the mean-field (or Hartree-Fock) approximation is not correct even in qualitative aspects if the system is one-dimensional [10]. To overcome such inherent difficulties in the mean-field approximation a bosonization similar to that of TOMONAGA [11] has been developed [12-14] by introducing (1) a linearization of the energy spectrum around the Fermi energy, and (2) a simplification of the matrix elements of the mutual interaction, as $v(q \approx 0) = J$ and $v(|q| \approx 2k_F) = -J$, where k_F is the Fermi momentum. In addition, CROSS and FISHER [14] invented a scheme based on such a bosonization of the fermion operators, but with the parameters in the resulting boson Hamiltonian adjusted so that they lead to exact known results of the spin-wave velocity and the critical exponent of the spin-spin correlation function. This adjustment is justified since the above procedures, (1) and (2), which make the bosonization possible, is strictly correct for fermions near the Fermi energy; as a consequence in the original Hamiltonian (3), all of the fermions are affected by the adjusted interaction, which is of the order of magnitude of the band energy.

The Hamiltonian of Cross and Fisher is given by

$$H = v_F \frac{2\pi}{L} \sum_{q>0} [\rho_1(q)\rho_1(-q) + \rho_2(q)\rho_2(-q)] + \frac{J}{N} \sum_{q} [\rho_1(q)\rho_1(-q) + \rho_2(q)\rho_2(-q) + 4\rho_1(q)\rho_2(-q)] , \quad (6)$$

where $L = Na$, $v_F = Ja$, and the $\rho_i(q)$'s are density fluctuation operators defined by

$$\rho_1(q) = \sum_{k>0} a^+_{1k+q} a_{1k} , \quad \text{etc.} , \quad (7)$$

with $i = 1$ (2) indicating the fermions with wave-vectors near k_F ($-k_F$). The $\rho_i(q)$'s satisfy the commutation relations

$$[\rho_1(-q),\rho_1(q')] = [\rho_2(q'),\rho_2(-q)] = \delta_{q,q'} qL/2\pi , \quad (8a)$$

$$[\rho_1(q),\rho_2(q')] = 0 . \quad (8b)$$

The availability of exact solutions for (6) together with (8a) and (8b) is clear, because the Hamiltonian is bilinear with respect to the boson operators, $\rho_i(q)$.

Although the scheme of Cross and Fisher is excellent, the important fact mentioned in Section 1 -- that there exist in the ground state large quantum fluctuations between magnetic and nonmagnetic spin configurations -- is not easily seen in their theory. On the other hand, the phase Hamiltonian approach, which was derived by NAKANO and FUKUYAMA [15,16] and uses another transformation beyond bosonization, is very convenient in this context. It reads as follows [15,16]:

$$H_0 = \int dx[A\,(\nabla\theta(x))^2 + C\,p(x)^2 - D\cos 2\theta(x)] , \quad (9)$$

with

$$[\theta(x), p(x')] = i\,\delta(x-x') , \quad (10)$$

where A, C, and D are defined by

$$A = \frac{Ja}{8} , \qquad C = \frac{\pi^2}{2} Ja , \qquad D = \frac{\pi^2}{8a} J . \quad (11)$$

The phase variable $\theta(x)$ is related to the z-component of the spin operators by

$$S^z(x) = \frac{1}{a}\cos(2k_F x + \theta(x)) + \frac{1}{2\pi}\nabla\theta(x) , \quad (12a)$$

i.e., on each lattice site $x = \ell a$,

$$S^z(\ell a) = S_\ell{}^z = \frac{(-1)^\ell}{a}\cos\theta(x) , \quad (12b)$$

if $\Theta(x)$ is uniform in space. The phase variable $\Theta(x)$ and its conjugate, $p(x)$, are naturally related to the $\rho_i(q)$'s in (7), as follows:

$$\Theta(x) = \Theta_+(x) \ , \quad p(x) = -\frac{1}{4\pi}\nabla\Theta_-(x) \ , \tag{13}$$

where

$$\Theta_\pm(x) = i\sum_{q\neq 0}(\rho_1(q) \pm \rho_2(q))\, A_q(x) \ , \tag{14}$$

$$A_q(x) = \frac{2\pi}{Lq}\exp\!\left(-\frac{\alpha}{2}|q| - iqx\right) \ , \tag{15}$$

and α is a convergence factor to be taken $\alpha \to 0$ whenever possible.

To be precise (6) and (9) are not equivalent, since the Umklapp scattering -- i.e. the last term in (9), the D term, which sometimes plays an important role -- is not taken into account in (6).

The phase Hamiltonian (9) was originally introduced by SUZUMURA [17] in another context to treat interacting one-dimensional electrons, where both charge and spin degrees of freedoms must be taken into account [18].

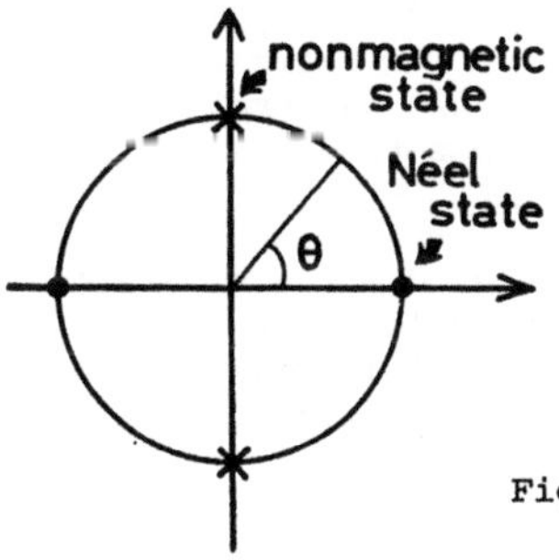

Figure 4. Classification of various states in the Θ-plane.

From (12b), we see that on the plane of the angle Θ, the values $\Theta = 0, \pm\pi$ correspond to the Néel state, and $\Theta = \pm\pi/2$ is the nonmagnetic state, as schematically shown in Figure 4. The term $\{D\cos 2\Theta\}$ in (9) appears to stabilize the Néel state, since $D > 0$. The quantum fluctuations represented by $\{C\, p^2(x)\}$, however, drastically reduce the effect of $\{D\cos 2\Theta\}$, resulting in essentially no contribution from this term. This cancellation can be seen by separating $\Theta(x)$ as

$$\Theta(x) = \Theta_s(x) + \hat{\Theta}(x) \ , \tag{16}$$

where $\Theta_s(x)$ is the slowly-varying classical part and $\hat{\Theta}(x)$ describes the quantum fluctuations around $\Theta_s(x)$. The simplest scheme to treat $\hat{\Theta}(x)$ is to work within the harmonic approximation

$$\cos 2\hat{\Theta} \to e^{-2\langle\hat{\Theta}^2\rangle}\,[1 - 2(\hat{\Theta}^2 - \langle\hat{\Theta}^2\rangle)] \ , \tag{17}$$

but with $\langle\hat{\Theta}^2\rangle$ to be determined self-consistently (in the so-called SCHA Self-Consistent-Harmonic-Approximation [15,18]). This scheme implies that the D-term in (9) is effectively reduced to $D \to D\exp[-2\langle\hat{\Theta}^2\rangle] \equiv \tilde{D}$; solution of the self-consistency equation leads to $\tilde{D} = 0$. Similar conclusions are obtained also by a renormalization group treatment [19-21]. Hence in the ground state $\Theta(x)$ cannot be pinned but fluctuates arbitrarily, which is to be as a manifestation of the fact stressed in Section 1 -- states with spin configurations very different from that of the Néel state, occupy an important fraction of the ground state.

4. One-Dimensional Spins with Lattice Distortions.

In the presence of the lattice distortion, u_ℓ, the original Hamiltonian can be modified as follows

$$H = \sum J(\ell,\ell+1)\, S_\ell \cdot S_{\ell+1} , \qquad J(\ell,\ell+1) = J(1 + \lambda \frac{u_\ell - u_{\ell+1}}{a}) , \tag{18}$$

where λ represents the dimensionless coupling constant between the spins and the lattice distortions. For dimerization, the situation of interest here, $u_\ell = u\,(-1)^\ell$, the phase Hamiltonian for (18) can be written as

$$H = \int dx\, [A\,(\nabla\Theta)^2 + C\, p^2 - D\cos 2\Theta - B\sin\Theta + \frac{2K}{a} u^2] , \tag{19}$$

where $B = J\lambda u/a^2$ represents the coupling between spins and the dimerization distortion, and the last term in (19) is the elastic energy of the dimerization. Now it is obvious that the B term favors the singlet state, $\Theta = \pi/2$. Actually, the energy gain, E_s, associated with the B term is found to be [15,22], again within SCHA,

$$E_s/L = -A\,(J/a)\,(\lambda u/a)^{4/3} , \tag{20}$$

where $A = 0.35$ if $D = 0$, and $A = 0.305$ if D is given by (11). Such a unique dependence on u -- $u^{4/3}$ -- was first obtained by CROSS and FISHER [14]. In more elaborate treatments based on the renormalization group, extra logarithmic dependences on u have been found [20,23]; but the exponent of 4/3 has been agreed upon by various investigations, including numerical calculations [24-26] which also test favorably the prefactor in (20).

Equation (20) together with the fact that the cost in energy due to the elastic distortion -- represented by the last term in (19) -- is proportional to u^2 (i.e. with an exponent of u higher than 4/3) leads to a conclusion that if the exchange interaction between spins is purely one-dimensional, there always exist finite dimerizations; i.e. the spin-Peierls state is stabilized.

Experimental results, however, indicate that the SP transition is not so commonly observed in reality, but the magnetic phase transition is far more usual. This observation points to the fact that the interchain exchange interaction is very important in actual systems.

5. Interchain Exchange Interactions. [27]

We consider the interchain exchange coupling given by

$$H' = -J' \sum_{\langle n,n'\rangle} \sum_\ell S_{n\ell} \cdot S_{n'\ell} , \tag{21}$$

where $S_{n\ell}$ is the spin at site ℓ of the n-th chain, and J' only couples neighboring chains. We assume that, in the plane perpendicular to the chain, the chain sites can be divided into two sub-lattices, e.g., a square lattice.

The effect of (21) can be reasonably treated by the mean-field approximation [28], which results in [27]

$$H_{3D} = H + \int dx\, [-F\cos\Theta + F/2\,\langle\cos\Theta\rangle], \qquad F = \langle\cos\Theta\rangle\, Z|J'|/a , \tag{22}$$

where H is given by (19), and Z is the number of nearest-neighbor chains. In the presence of F the stable configurations are found to be either Néel state or SP state, and the resulting phase diagram is shown in Figure 5, where j' and η are

$$j' = Z|J'| / \pi^2 J , \qquad \eta = J\lambda^2 / 4\pi^2 K a^2 . \tag{23}$$

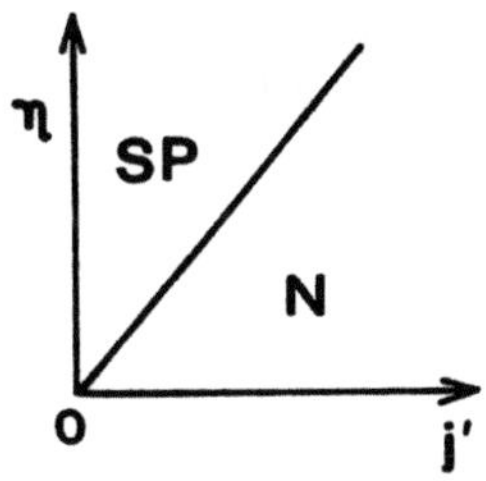

Figure 5. Competition between the Néel (N) and the spin-Peierls (SP) states.

From Figure 5 it can be seen that a small K is favorable to the SP state, and this is the reason why the SP state is observed only in the case where the lattice is softened at the particular wave-vector corresponding to the dimerization [7,29,30].

It is interesting to note [27] that there are qualitative differences in the competition between the N and the SP states depending (in a broad sense) on the type of such perturbations as the Ising anisotropy and the staggered field in a single chain. It is to be inferred from the above that each perturbation picks up different quantum fluctuations of the ground state of one-dimensional Heisenberg antiferromagnets.

6. Magnetic Fields. [31]

In the presence of a uniform magnetic field, B, we have the Zeeman energy given by

$$H_z = -g\mu_B B \sum_{\ell} (a_\ell^{\dagger} a_\ell - 1/2) = -g\mu_B B/2\pi \int dx\, \nabla\theta \, . \tag{24}$$

The possible phase diagram in the field is shown in Figure 6 [31], where j' and η are defined by (23) and $h \equiv g\mu_B B/\pi^2 J$. In this figure, SF and MD are the spin-flop and the magnetic discommensurate states respectively, each of which is schematically shown in Figure 7. Especially the MD state is a lattice of spin 1/2 solitons in a field of dimerized singlets. If the distance between solitons is large compared to their width, the spatial variation of Θ_s defined by (16) is given by

$$\cos\Theta_s = \pm \tanh \frac{x}{\xi} \, , \qquad \xi = 3\pi K a^3 (1-j')^{1/2} \,/\, J\lambda^2 (1-3j'/2\eta)^{1/2} \, . \tag{25}$$

The existence of such a soliton lattice induced by magnetic fields has also been noted in numerical calculations by HARADA and KOTANI [32], who treated a single chain within the Hartree-Fock approximation.

7. Experiments.

In this section we briefly discuss some particular aspects of the phase diagram. By measuring the magnetization of TTF-CuBDT, BLOCH et al. [33] observed a

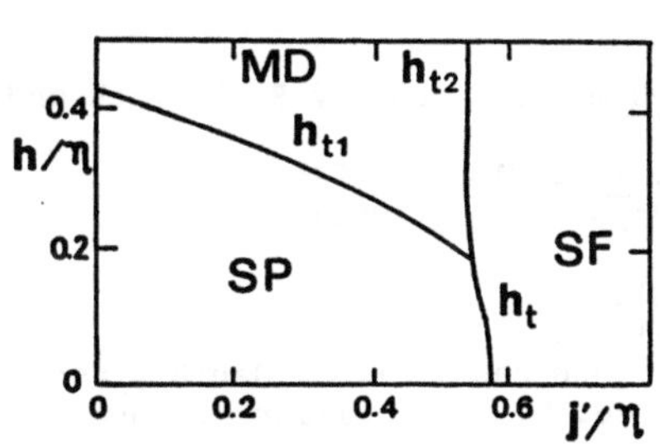

Figure 6. Phase diagram in a uniform magnetic field.

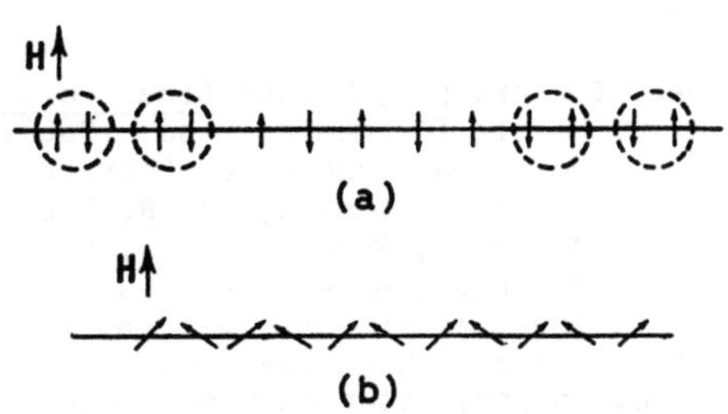

Figure 7. (a) Magnetic discommensurate (MD) state; (b) spin-flop (SF) state.

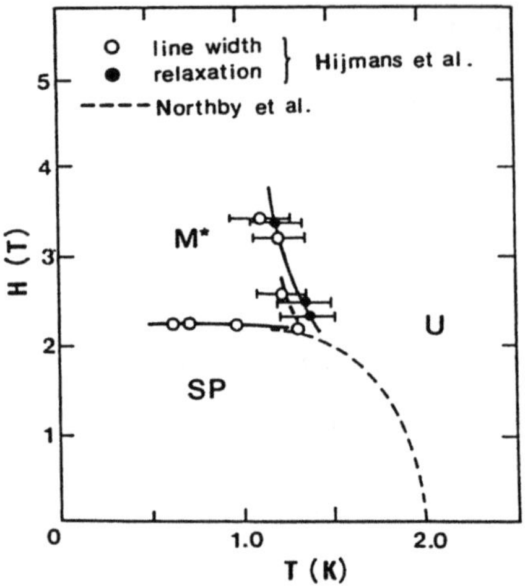

Figure 8. Phase diagram in the magnetic field temperature plane of TTF-AuBDT (after HIJMANS et al. [34]).

first-order transition as the magnetic field was varied at low temperatures. Recently HIJMANS et al. [34] obtained by Nuclear Magnetic Resonance the results of Figure 8 on TTF-AuBDT, which behaves similarly to TTF-CuBDT but with a lower magnetic field for the onset of the SP → M* transition at low temperatures. Through the analysis of the line shape of echos, HIJMANS et al. conclude that this M* state can indeed by identified with the MD state discussed in Section 6.

8. Summary.

We first demonstrated that the phase Hamiltonian is a powerful tool to treat the large quantum fluctuations intrinsic to one-dimensional S = 1/2 Heisenberg antiferromagnets. Based on this phase Hamiltonian, we examined the three-dimensional effects present in actual solids; i.e., the spin-lattice coupling and the interchain exchange interaction. We found these two interactions severely compete with each other resulting in either a spin-Peierls or the Néel state. We also examined the effects of magnetic fields in the presence of such three-dimensional interactions.

The interesting features of quasi-one-dimensional Heisenberg models are obviously far from having been completely explored. Although there already exist some investigations [35,36], finite temperatures should be examined in detail and comparison with experiments carried out. The non-adiabaticity of the lattice vibrations should also be of interest in some cases.

We acknowledge a Grant-in-Aid for a Special Project Research on the Properties of Molecular Assemblies from the Ministry of Education, Science and Culture.

References

1. See, for example, Mathematical Physics in One-Dimension, ed. by E. H. Lieb and D. C. Mattis (Academic Press, 1966).
2. J. des Cloizeaux and J. J. Pearson, Phys. Rev. 128, 2131 (1962).
3. T. Nagamiya, K. Yosida, and R. Kubo, Adv. in Phys. 4, 97 (1955).
4. P. W. Anderson, Phys. Rev. 86, 694 (1952).
5. P. W. Anderson, Mater. Res. Bull. 8, 153 (1973).
6. L. Hulthèn, Arkiv. Mat. Astron. Fysik 26A, 1 (1938).
7. For an excellent review, see J. W. Bray, L. V. Interrante, I. S. Jacobs, and J. C. Bonner, Extended Linear Chain Compounds, ed. by J. S. Miller (Plenum Press, 1983), Vol. III.
8. J. C. Bonner and M. E. Fisher, Phys. Rev. 135, A640 (1964).
9. P. Jordan and E. P. Wigner, Z. Physik 47, 631 (1928).

10. See, for example, J. Solyom, Adv. in Phys. 28, 201 (1979).
11. S. Tomonaga, Prog. Theor. Phys. 5, 349 (1950).
12. A. Luther and I. Peshel, Phys. Rev. B 9, 2911 (1974); ibid 12, 3908 (1975).
13. A. Luther and V. J. Emery, Phys. Rev. Lett. 33, 589 (1974).
14. M. C. Cross and D. S. Fisher, Phys. Rev. B 19, 402 (1979).
15. T. Nakano and H. Fukuyama, J. Phys. Soc. Jpn. 49, 1679 (1980).
16. T. Nakano and H. Fukuyama, J. Phys. Soc. Jpn. 50, 2489 (1981).
17. Y. Suzumura, Prog. Theor. Phys. 61, 1 (1979).
18. For a general review, see H. Fukuyama and H. Takayama, Electronic Properties of Inorganic Quasi One Dimensional Compounds, ed. by P. Monceau (D. Reidel, 1984), p. 41.
19. F. D. M. Haldane, Phys. Rev. Lett. 45, 1358 (1980).
20. J. L. Black and V. J. Emery, Phys. Rev. B 23, 429 (1981).
21. M. P. M. den Nijs, Phys. Rev. B 23, 6111 (1981).
22. S. Inagaki and H. Fukuyama, J. Phys. Soc Jpn. 52, 2504 (1983).
23. G. L. M. Dassen and M. A. de Moura, Phys. Rev. B 30, 467 (1984).
24. W. Duffy, Jr. and K. P. Barr, Phys. Rev. 165, 647 (1968).
25. J. C. Bonner and H. W. Blöte, Phys. Rev. B 25, 6959 (1982).
26. K. Okamoto, H. Nishimori, and Y. Taguchi, submitted to J. Phys. Soc. Jpn.
27. S. Inagaki and H. Fukuyama, J. Phys. Soc. Jpn. 52, 3620 (1983).
28. D. J. Scalapino, Y. Imry, and P. Pincus, Phys. Rev. B 11, 2042 (1975).
29. D. E. Moncton, R. J. Birgeneau, L. V. Interrante, and F. Wudl, Phys. Rev. Lett. 39, 507 (1977).
30. B. van Bodegom, B. C. Larsen, and H. A. Mook, Phys. Rev. B 24, 1520 (1981).
31. S. Inagaki and H. Fukuyama, J. Phys. Soc. Jpn. 53, 4386 (1984).
32. I. Harada and A. Kotani, J. Phys. Soc. Jpn. 51, 1737 (1982).
33. D. Bloch, J. Voiron, J. C. Bonner, J. W. Bray, I. S. Jacobs, and L. V. Interrante, Phys. Rev. Lett. 44, 294 (1980).
34. T. W. Hijmans, H. B. Brom, and L. J. de Jongh, Phys. Rev. Lett. 54, 1714 (1985).
35. J. Kondo, Physica 123B, 169 (1984).
36. S. Inagaki and H. Fukuyama, J. Magn. Magn. Mater. 54-57, 1261 (1986), and to be published.

Static and Dynamic Properties of Random Exchange (S=1/2) Heisenberg Chains

J.V. José

Department of Physics, Northeastern University, Boston, MA 02115, USA

This paper reviews briefly the thermodynamic and dynamic properties of charge transfer compounds like $Qn(TCNQ)_2$. These are systems that are described in terms of an isotropic Heisenberg model with antiferromagnetic random exchange and $S = 1/2$. A renormalization-group approach to the problem is shown to lead to universal results, i.e. the thermodynamic properties do not depend on the microscopic details of the interactions between spins. Experimental evidence supporting the universality result is discussed. A dynamic treatment of this problem is much more complicated. However, a dynamic study of a random-exchange-antiferromagnetic-Ising model leads to results that correlate positively with some current experimental results. A critical view of the present situation between experiment and theory of the equilibrium and nonequilibrium properties of these systems is also presented.

1. Introduction

Physical studies of lower dimensional systems have been the source of great interest in recent years. This surge of activity has been due in part to important innovations in the fabrication of new materials that have highly anisotropic properties. Good examples of this trend are the studies of organic quasi-one-dimensional conductors. In most cases organic solids are insulators. However, in charge-transfer complexes the anisotropic conductivities can be as large as in metals at room temperatures. A stable charge-transfer complex is usually formed by the partial transfer of an electron from a donor to an acceptor molecule forming an ionic crystal. Extensively studied examples of these salts are combinations that include tetracyano-p-quinodimethene (TCNQ). Most experimental and theoretical studies have been directed at understanding the conducting properties of these salts. In 1972 BULAEVSKII et al. (BZKLS hereafter) discovered that in some of these salts the low temperature magnetic equilibrium properties exhibit singular behavior [1]. One of the most studied specimens that shows this behavior is the quinolinium-$TCNQ_2$, or $Qn(TCNQ)_2$ salt [1,2,3]. This is a charge-transfer salt as well, but its transport properties differ from those of the pure TTF(TCNQ), which has also been studied extensively. In particular, its conductivity decreases sharply as the temperature is lowered. From X-rays and recent NMR experiments, it is believed that the decrease of conductivity is caused by increased localization due to disorder. The disorder is produced by the electric dipole moment of the Qn chains that can have two possible orientations at random [4,5]. BZKLS found that their experimental results could not be explained by standard theories that did not include disorder and the quantum nature of the spins. They proposed a random-exchange-Heisenberg-antiferromagnetic-chain model (REHAC) with spin 1/2, to explain their experiments. Further detailed experiments have extended the results found by BZKLS including their dynamic properties [3].

It was only in recent years that an understanding of the anomalous equilibrium properties of these salts became possible. Three different groups, almost simultaneously, provided a renormalization-group type analysis that explains these anomalous results [6,7,8]. All three approaches use, in one way or another, the idea of thinning of degrees of freedom. Here we will discuss the approach given in Reference 8, since this is the one the author understands best. With regard to the dynamic properties, there is some evidence that these systems show slow decay of the recovery magnetization [3]. Here the theoretical studies are less complete. Studies of the dynamics of a REHAC are highly complicated and even direct numerical approaches are incomplete. A phonomenological approach that fits the logarithmic behavior of the linewidth has appeared [9]. A more microscopic analysis, at the same level of rigor as that for the static properties of the REHAC, would be desirable. Since a dynamic analysis of the REHAC is highly complicated, it was decided [10] that a study of a random-exchange- Ising-chain, which is difficult in itself, could shed some light onto the more complicated dynamics of a REHAC. It was found that, in fact, some of the results from this model do match some of the experimental findings. This correspondence is incomplete and more experimental and theoretical studies on these systems are required before a theory of the dynamic properties can be made to correspond to all the experimental findings.

In this contribution I emphasize the main ideas about the characteristics of these salts and the theoretical approaches that have led to some understanding of their properties. In Section 2, a brief review of the experimental magnetic properties of $Qn(TCNQ)_2$ is given. In Section 3, theoretical approaches prior to the renormalization group (RG) treatments are discussed. In Section 4, the main ideas of the RG treatment are discussed from a general standpoint. Section 5 deals with the dynamics of the random-exchange-Ising model described by a Glauber model. Finally, in Section 6 we present a critical view of the present experimental and theoretical situation about the physical behavior of these materials. A more detailed review of the points discusseed in this paper will appear elsewhere [11].

2. Experimental Results

One of the main characteristics of a REHAC is that its zero field magnetic susceptibility, χ, shows a divergent behavior for temperatures from around 20 K to about 7 mK. Below this temperature, the three- dimensional nature of the specimens takes over [12]. Experimentally, it is found that the exchange between localized spins along the TCNQ chains is antiferromagnetic and isotropic. Specifically, it is found that

$$\chi \sim T^{-\alpha_\chi} \quad . \tag{1}$$

The α_χ's take values between 0.7 and 0.9. Other magnetic properties show unusual behavior as well, i.e. the magnetic specific heat behaves like

$$C \sim T^{(1-\alpha_C)} , \tag{2}$$

while the high-field magnetization and specific heat behave like $M \sim H^{(1-\alpha_M)}$ and $C \sim TH^{-\alpha_C}$, respectively. Here H is the externally applied magnetic field. Within the experimental accuracy of the measurements, it is found that the exponents α_χ, α_C, and α_M are the same. Also, the measurements where (1) is valid cover several decades in temperature, and the exponents are essentially temperature independent [1-3]. The aim of the theory is to explain the power law behavior found in the magnetic properties of these specimens.

With regard to the dynamic properties of these systems, less is known. There are cw ESR experiments that show a logarithmic behavior of the linewidth as a function of temperature [3,9]. The property of interest here is the behavior of the recovery magnetization, M, seen in pulsed ESR experiments [3]. It shows a fast exponential decay in time, followed by a non-exponential decay M_r that takes a long time to relax to its equilibrium value. These experiments are done in the lower end of the temperature scale (below 0.1 K) where the REHAC behavior is dominant. This type of decay is reminiscent of usual spin-glass systems. The experimental results are not sufficiently clear yet to provide an explicit form for the M_r as a function of time.

3. Early Theoretical Treatments

In the original paper by BZKLS, it was suggested that the appropriate model to describe the behavior of the charge transfer salts like $QN(TCNQ)_2$ has the Hamiltonian

$$H = \sum_m [J_m \vec{S}_m \cdot \vec{S}_{m+1} - h S^z_m] \quad . \tag{3}$$

Here, $\vec{S}_m$ is the usual Pauli spin 1/2 operator at site m. The last term of (3) is the Zeeman energy with $h = g\mu_B H$, where g is the Lande factor, and μ_B is the Bohr magneton. In the case where the exchange constant J_m is the same along the chain, H represents a periodic model. This has been studied numerically. It is found that in the periodic case, the magnetic susceptibility has a maximum as a function of T: it decreases and tends to a constant value as T goes to zero [1]. This fact led BZKLS to suggest that, in fact, the $\{J_m\}$ are random. To calculate the thermodynamic properties defined by the Hamiltonian of (3), we need to calculate the quenched averaged free energy associated with H. This is given by $[\beta F] = -[\ln Z]$ with $Z = \mathrm{Tr}\, e^{-\beta H}$ the configurational partition function, and the brackets [,] stem for configurational average with respect to the probability law $P(J_m)$. The problem is then well posed whenever we provide a specific form for $P(J_m)$.

Since BZKLS did not know $P(J_m)$, they resorted to an alternative heuristic argument. The Hamiltonian H can be separated into two components: an x-y component and an Ising-like component. When J is uniform, the x-y component maps into a spinless free electron system that can be solved exactly. The effect of adding the Ising component to H introduces interactions between the electrons, and these interactions are strong. BZKLS made the assumption that this interacting gas of fermions could be replaced by a "renormalized quasi- particle-glass", in the spirit of the Landau theory of Fermi liquids. With this assumption, they constructed an appropriate form for the density of states for this glass that would fit the experiments. The fits to their experiments seemed to be satisfactory.

Several authors criticized the BZKLS theoretical approach. THEODOROU and COHEN (TC) tried a more microscopic treatment of the problem by starting with a Hubbard model with random onsite energy [13]. They were able to show that the H given in (3) does emerge from a perturbative calculation at low temperature. More importantly, they were able to derive a probability distribution for $P(J_m)$. They found that P(J) behaves like

$$P(J) \sim J^{-\alpha_{TC}} \quad . \tag{4}$$

This probability law is singular for small J's, with $0 \leq \alpha_{TC} < 1$; equivalently small values of J are dominant in this type of probability law. From the TC analysis, the problem is completely defined. We should, in principle, proceed to evaluate the quenched partition function defined by H and P(J). This is, however, a highly complicated problem. Suffice just to say that even the periodic case is non-trivial, even at zero temperatures. Thus, if progress is to be made, approximate schemes should be looked for.

THEODOROU and COHEN [13] suggested that the divergence in χ should be directly related to the divergence of P(J), i.e. P(J) with no divergence for small J's should not lead to a divergent χ. TIPPIE and CLARK [3] studied the TC model directly. They made the drastic approximation of considering only two spins at a time, however, their results were essentially indistinguishable from those of BZKLS. In these analyses the exponent in χ is the same as that of the probability law, and thus temperature dependent.

4. Renormalization Group Analysis

The BZKLS, TC, and Tippie and Clark approaches provide qualitative explanations of the experimental results. However, when one tries to improve these schemes, the results change significantly. An explicit numerical evaluation of [F] for four spins leads to a divergent χ, but with an exponent that acquires a logarithmic correction in temperature [8]. This result differs from the approximate treatments discussed above. Specifically, all these approaches reduce the many-body problem defined by H and P(J), to that of an ensemble of noninteracting dimers. Looking at the four-spin problem indicates that there are important renormalizations due to the many-body nature of H. With the aim of gaining a deeper understanding of the physics of this model, three different groups of researchers introduced methods to treat the many-body interactions more accurately [6-8]. A common feature of the three different treatments is that they entail an iterative elimination of the degrees of freedom in the model. Here I discuss a method that was introduced by HIRSCH and the author [8]. The method is closer to the usual RG approach, save for important extensions that were introduced to treat the REHAC. The results from the three different approaches are, for the most part, equivalent.

First, the chain is divided into cells of three spins. Taking an odd number of spins in each cell is important to map a S = 1/2 Hamiltonian to a S = 1/2 renormalized H. The H for the cell is diagonalized exactly. Let us denote the three spins in the cell as 1, 2, 3. The coupling between 1 and 2 is J_1 and between 2 and 3 is J_2. This cell interacts with a near neighbor cell with spins 4, 5, and 6 with interactions J_4 and J_5, respectively. The interaction between the cells is J_3. The RG procedure entails replacing the cell by a renormalized spin, with a renormalized coupling constant between cells. After the RG procedure, we get a new coupling J'_3 between the new cells. Evaluation of J'_3 via lowest order of perturbation theory in J_3 gives $J'_3 = J_3A(J_1,J_2)A(J_5,J_4)$. To a good approximation $A(J_1,J_2) \sim J_1J_2$. This implies that $J'_3 \sim J_3J_1J_2$. This multiplicative nature of the RG is essential to the physics of the REHAC. Since we are dealing with a random problem, the J's are chosen from the distribution P(J). As we renormalize the J's, this probability law gets renormalized as well. The renormalized P(J) is calculated from the expression

$$P'(J'_3) = \int[\Pi_i dJ_i P(J_i)]\delta(J'_3 - J_3A(J_1,J_2)A(J_5,J_4)).$$

Thus, we have a chain of renormalized J's that leads to a chain of renormalized P(J)'s. The remarkable feature about the RG chain is that, because of the multiplicative nature of the RG equations for the J's, the renormalized P(J)'s show a universal behavior. What this means is that, provided J is bounded, after a sufficient number of RG iterations, the P(J) will reach a universal form that will not depend on the initial P(J). The universal form for P(J) is of the type given in (4). However, the exponent α is J dependent. Thus, regardless if we start with a P(J) that is or not singular, the renormalized P(J) will be singular.

The result presented above was derived from RG analysis at T = 0. To obtain the experimental consequences of this result, one has to extend the analysis to finite temperatures. One way to do this is to include the higher-lying excitations of the cells and their interactions in the evaluation of [F]. A lowest order approximation that essentially divides the chain between regions where the low-lying or high-lying excitations dominate, leads to a calculation of the thermodynamic properties of a REHAC. The resulting forms for χ and C have the forms given in (1) and (2). However, an important difference is that the exponents are different for different quantities. More importantly, these exponents have a temperature dependence. This dependence, although only logarithmic, should be seen experimentally, since the experiments cover several decades in temperature. BONDESSON and SOOS [7,9] have made explicit comparisons to experiment. They find that instead of starting with a continuous P(J), they get a better fit to the experiments by taking a P(J) of the form $P(J) = p\delta(J_0\varepsilon - J) + (1-p)\,\delta(J_0 - J)$. The same RG discussed above can be implemented for this type of P(J), and it is found that the renormalized P(J) of the form given in (4) is also obtained [14].

5. Dynamics in a REHAC

As mentioned before, we are interested in gaining some understanding of the dynamics of a REHAC. However, since it is unclear how to study the REHAC dynamics at the same level of rigor as in the static case, we resorted to the study of a simpler system, keeping some of the properties of the real REHAC problem: the randomness of the exchanges, and the corresponding divergence of the χ as given in (1). The model studied is that of Glauber with random exchanges [15]. The thermally averaged onsite magnetization, q_i, satisfies the equation of motion,

$$\frac{dq_i}{dt} = -q_i + \Gamma_i^{+} q_{i+1} + \Gamma_i^{-} q_{i-1} \ . \tag{5}$$

The $\Gamma^{\pm}$ are the hopping matrices, with the symmetry property
$\Gamma^{+}(J_i,J_{i-1}) = \Gamma^{-}(J_{i-1},J_i)$.
This asymmetry relation satisfied by all the Γ's is very important in the study of the dynamics of this model. This asymmetry makes the problem more complicated, however. To correlate with the experimental measurements of the pulsed-ESR experiments, one is interested in calculating the quenched average magnetization $M = [M(J,t)]_J$. As before, the $[,]_J$ means average with respect to P(J). To evaluate M, the continued fraction method is employed to calculate the Green's function associated with the equation of motion given in (5). There it is found that M has two components: a rapid exponential decay, and a remanent slow component that goes like [10],

$$M_r \sim 1/3 - f(\alpha)[T \ln t] - g(\alpha)[T \ln t]^{3(1-\alpha)} + \dots \ . \tag{6}$$

Here $T \ln t \ll 1$. The $T \ln t$ term dominates for values of $0 \leq \alpha < 0.5$, while the $[T \ln t]^{3(1-\alpha)}$ term dominates for $0.8 \leq \alpha < 1$. This logarithmic decay of the remanent magnetization is seen often in spin- glasses. It is of importance to mention, however, that this type of remanence does not exist if the probability law is discrete rather than continuous. The remanence in M_r decreases as the temperature increases, in agreement to what is seen experimentally [3]. The physical explanation for the two time decays is as follows: Because of the particular form of P(J) we can divide the chain into clusters of strongly coupled spins separated by weaker couplings. Let the initial state be the one where all the spins point in the same direction. When the system is allowed to relax to its equilibrium state, in the absence of a magnetic field, two things happen. First, the spins inside the clusters relax to their minimum state quickly. For an Ising chain with antiferromagnetic couplings, the lowest energy state is the one with zero total magnetization. After the initial intracluster relaxation, the total chain magnetization is not zero. This is because there are two possible intracluster equilibrium states, and these will not be the same for all clusters after the initial relaxation. For the whole chain to reach its lowest energy state, whole clusters of spins have to flip at once, to lower the total magnetization. This flip has a large cost in energy, and it will take a long time to occur, since the intercluster couplings are weak. The specific details as to how this decay takes place and how it is directly related to the asymmetry relation of the Γ's is given in Ref. 10, where other quantities, like hysteresis loops, are discussed.

Although these results were obtained in detail for an Ising model, it is possible that all one-dimensional disordered magnetic systems show this type of remanent behavior.

6. Conclusions and Outlook

I have reviewed briefly the experimental and theoretical situation with regard to the magnetic and dynamic properties of a REHAC. There is theoretical agreement that the model defined in (3), with the J_m anarbitrary but bounded, random variable, has thermodynamic properties which are independent of the P(J) chosen, and of the form of (1) and (2). This universality behavior has been confirmed experimentally [16]. On the other hand, the logarithmic corrections in temperature for the exponents α do not agree with the experimental results that cover up to three decades in temperature. It is possible that the REHAC does not give a full description of the real $Qn(TCNQ)_2$ specimens [17].

As to the dynamic properties, the situation is much less developed. Of course, dynamic studies are always more difficult than static ones. There are some experimental results of cw and pulsed ESR studies in different frequency and temperature regimes. Here I discussed a microscopic study of the dynamics of a random Ising-Glauber model, with similar properties to that of a REHAC. The essential result from this study is that there are strong remanent properties in the magnetization. This seems to agree with the ESR pulsed experiments. Since the correct model to describe the dynamics of $Qn(TCNQ)_2$ is a REHAC, there are two possible explanations as to why the Ising results seem to agree qualitatively with the experiments. Since the remanence is observed at very low temperatures, it is not unlikely that the Hamiltonian representing the real samples have a small ion anisotropy that could make the Ising model more appropriate. More likely is the fact that this remanent behavior is a universal property of a disordered Hamiltonian model.

Acknowledgments

I want to thank my collaborators in most of this work, J. T. Choi, J. Hirsch, M. Mehl, and J. B. Sokoloff. Also, many conversations with G. W. Clark have been illuminating. This work supported in part by U.S. National Science Foundation through grant DMR 85-00035.

References

1. L. N. Bulaevskii, A. V. Zvarykina, Y. S. Karimov, R. B. Lybovskii, and I. F. Shegolev, Zh. Eksp. Teor. Fiz. 35, 725 (1972) [Sov. Phys. JETP 35, 384 (1972)].
2. Reviews of the experimental properties of $Qn(TCNQ)_2$ can be found in: W. G. Clark, Physics in One Dimension, edited by J. Bernasconi et al. (Springer-Verlag, 1981), p. 289; W. E. Hatfield and L. W. ter Haar, Ann. Rev. Mater. Sci. 12, 177 (1982).
3. L. C. Tippie and W. G. Clark, Phys. Rev. B23, 5846 and 5854 (1981).
4. I. F. Shegolev, Phys. Stat. Solidi 12, 9 (1972).
5. A. M. Vainrub et al., JETP Lett. 41, 571 (1985).
6. S. K. Ma, C. Dasgupta, and C. K. Hu, Phys. Rev. Lett. 43, 1434 (1979), and Phys. Rev. B22, 1305 (1980).
7. S. R. Bondesson and Z. G. Soos, Solid State Comm. 35, 11 (1980), and Phys. Rev B22, 1793 (1980).
8. J. E. Hirsch and J. V. José, J. Phys. C 13, L53 (1980), and Phys. Rev. B22, 5339 (1980).
9. Z. G. Soos and S. R. Bondesson, Solid State Comm. 39, 289 (1981).
10. J. V. José, M. Mehl, and J. B. Sokoloff, Phys. Rev. B25, 2026 (1982), and B27, 334 (1983).
11. J. V. José, private communication.
12. H. M. Bozler, C. M. Gould, and W. G. Clark, Phys. Rev. Lett. 45, 1303 (1980).
13. G. Theodorou and M. H. Cohen, Phys. Rev. Lett. 37, 1014 (1976).
14. J. T. Choi and J. V. José, private communication.
15. R. J. Glauber, J. Math. Phys. 4, 294 (1963).
16. J. Sanny, G. Grüner and G. W. Clark, Solid State Comm. 35, 657 (1980).
17. J. Hirsch and R. Kariotis, private communication.

Part VII

Quasi-Two-Dimensional Systems

Two-Dimensional Magnetism in Graphite Intercalation Compounds

M.S. Dresselhaus

Massachusetts Institute of Technology, Cambridge, MA 02139, USA

The possibility of preparing single isolated magnetic layers either by intercalation or by atomic deposition processes has stimulated a great deal of experimental activity in an effort to study 2-dimensional magnetic phenomena. A brief review is given of efforts to study 2D magnetism in graphite intercalation compounds (GICs).

1. Introduction

The most promising experimental access to 2-dimensional (2D) magnetism in bulk samples is provided by superlattices prepared by atomic deposition processes or by intercalation. Complementary information is provided by superlattices synthesized by each of these techniques, each offering advantages and disadvantages. The greatest advantage of the atomic deposition process relates to the flexibility in the choice of materials and periodicities. The greatest control of the layer thickness and in-plane homogeneity is provided by molecular beam epitaxy (MBE), though metal organic chemical vapor deposition (MOCVD) offers almost as much control and greater sample throughput. With MBE, it is possible to prepare a superlattice from layers of magnetic species A and non-magnetic species B, or even from layers of more than one magnetic species. Layers of almost any magnetic element or compound can be deposited, thereby allowing in principle the study of 2D ferromagnetic, antiferromagnetic, ferrimagnetic, and spin glass behavior. Likewise, the non-magnetic species can be chosen to be insulating, semiconducting, or metallic. With MBE, the layer thickness d_A and d_B can be controlled over wide ranges (5 $Å \leq d \leq 200$ Å), and samples with several hundred superlattice unit cells $d = d_A + d_B$ can be fabricated (see Figure 1.).

The special advantage of intercalation compounds for studying 2D physics relate to their unique structural features, allowing preparation of ordered isolated magnetic layers. Intercalation compounds in general are formed by the insertion of atomic or molecular layers of a guest chemical species, called the intercalant, between layers of a host material. To allow insertion of layers of guest species, the host material is required to be a layered material, with very strong in-plane bonding and very weak interplanar bonding. There are many examples of host

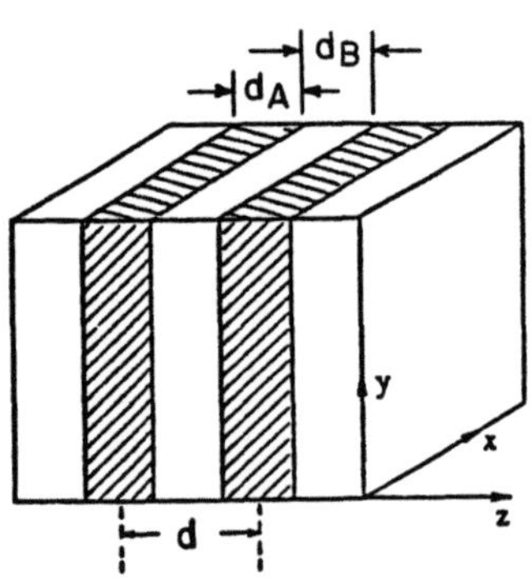

Fig. 1. Metallic superlattice with constituents M_A and M_B having layer thicknesses d_A and d_B where $d_A + d_B = d$.

materials which support intercalation, including graphite [1,2,3] (which has been studied most extensively), transition metal dichalcogenides [4], some clays [5], and some polymers and gels [6]. The basic reason why intercalation compounds are especially well suited for studying 2D physics relates to the staging phenomenon, by which n layers of a host material are sandwiched between sequential intercalate layers, where n is the stage index. The staging phenomenon thus gives rise to a periodic array or superlattice of intercalate sheets (a 1D superlattice as in Figure 1), analogous to the atomic deposition methods. Staging allows control of the separation between sequential intercalate layers, thereby providing a means for control of the interplanar exchange interactions, a feature exploited in studies of 2D magnetism.

The unique advantage of intercalation compounds with regard to studies of 2D magnetism is the atomically sharp interfaces that can be achieved between materials A and B and the large (macroscopic) distances over which these sharp interfaces are maintained [7]. Thus it is possible to accurately control the interlayer exchange interaction J' by intercalation; this level of control of J' is not possible by any of the atomic deposition processes. Since the magnetic behavior of atomically sharp isolated monolayers is qualitatively different from that of multiple magnetic layers [8], intercalation compounds provide a unique opportunity for experimental study of the magnetic properties of isolated magnetic layers (in high stage compounds) and of the effect of weak interplanar exchange coupling on these properties (as the stage index is decreased).

On the other hand, there are no known intercalation hosts which permit both the insertion of more than one consecutive magnetic layer and at the same time exhibit an ordered superlattice sequence in the direction normal to the layers. While graphite intercalation compounds (GICs) yield the highest degree of staging perfection (i.e., the best superlattice ordering), they only permit the insertion of single magnetic layers. In contrast, clay hosts permit the insertion of multiple consecutive magnetic layers, but clay intercalation compounds do not exhibit a high degree of staging order [5]. Also, there is only a limited choice of host materials and magnetic species that have been successfully intercalated, in contrast to the wide choice of constituents A and B possible with MBE. However, both donor (i.e., where electrons are transferred to the host material) and acceptor (i.e., where electrons are received from the host) magnetic GICs can be synthesized, though it is only for the acceptors that high stage compounds have been prepared.

For GICs, the c-axis repeat distance I_c between sequential intercalate layers follows the simple relation $I_c = d_s + (n-1)c_o$ where c_o is the interlayer separation in the pristine graphite host material and d_s is the separation of two host layers between which the intercalant is sandwiched. Typical values of d_s for magnetic intercalants are 4.85 A and 9.30 A for the Eu and $CoCl_2$ intercalants, respectively. Magnetic intercalation compounds with stages as high as 6 have been synthesized, corresponding to $I_c \approx 26$ A. For the layered magnetic materials normally investigated, the interplanar exchange interaction J' is small compared with that for the intraplanar interaction J. Assuming dipolar interactions, the J'/J ratio is further reduced by an order of magnitude by intercalation to a stage 2 GIC and by nearly 2 orders of magnitude for stage 6. Even for stage 1 GICs, highly anisotropic magnetic properties are observed.

A second major advantage of GICs for the study of 2D magnetism is the high degree of in-plane order that can be achieved. For typical magnetic acceptor GICs the magnetic intercalate layer is incommensurate with respect to their adjacent graphite layers; furthermore, the intercalate layers exhibit lattice constants identical with their values in the parent bulk compounds to within parts per thousand [2,3,9]. Since the in-plane intercalate lattice constants are almost identical to their bulk values, so are the in-plane magnetic exchange interactions. A priori knowledge of the in-plane exchange interactions in the magnetic superlattices leads to a major simplification in the interpretation of 2D magnetic phenomena. In contrast, magnetic donor compounds typically are commensurate with the graphite layers; for example, the rare earth C_6Eu compound exhibits a $(\sqrt{3} \times \sqrt{3})$ in-plane superlattice with respect to the graphite layers. The high degree of in-plane ordering in the magnetic donor compounds also leads to simplifications in the analysis of the magnetic properties, though in this case, all exchange interactions must be determined from the observed magnetic properties.

Synthesis techniques for the atomic deposition process for metallic superlattices are not sufficiently developed to form a crystalline in-plane structure with long range order. In contrast, GICs have been prepared with in-plane crystalline coherence distances in excess of 1 µm [10]. Crystalline in-plane ordering can, however, be achieved in diluted magnetic semiconductor (DMS) superlattices, as for example those based on $Cd_{1-x}Mn_xTe$. In this case, strained layer superlattices are formed, so that the lattice constants are weakly dependent on the relative layer thicknesses d_A/d_B. Because of the substitutional nature of the dilute magnetic cations, the magnetic interactions in DMS superlattices are mainly random and spin-glass ground states are observed.

2. 2D Magnetism Studies with GICs

By varying the separation between sequential magnetic layers through the intercalation mechanism, the interplanar magnetic coupling in graphite intercalation compounds can be reduced in a controlled manner to very small values, thereby providing a convenient system for studying 2D magnetism and the cross-over between 3D and 2D magnetic systems. In addition, by proper choice of intercalant, the spin dimensionality can be varied, as can be seen by considering the magnetic Hamiltonian

$$H = -J\sum_{(i,j)}\vec{S}_i \cdot \vec{S}_j - J_A \sum_{(i,j)}S_{iz}S_{jz} -J'\sum_{(i,k)}\vec{S}_i \cdot \vec{S}_k - J_A' \sum_{(i,k)}S_{iz}S_{kz} \quad (1)$$

in which J and J_A are the in-plane exchange and anisotropy energies and J' and J_A' are the inter-plane exchange and anisotropy energies. By increasing the stage index, the interplanar exchange and anisotropy terms J' and J_A' can be made sufficiently small so that the spins effectively form a 2D spatial system of isolated planes of magnetic spins. The exchange coupling between 3-dimensional spins $\vec{S}$ on a single magnetic plane has several possibilities for this 2D spatial system. When the anisotropy term J_A is small, then the exchange couplings are essentially isotropic and a Heisenberg spin system results; an example of such a system is the $FeCl_3$-GIC system. If, on the other hand, J_A and J have opposite signs and are of comparable magnitude, then only the x and y spin components contribute to (1) and the system is described by an XY model whereby the spins lie in the magnetic plane; an example of a 2D-XY system is the $CoCl_2$-GIC or the $NiCl_2$-GIC system. Finally, if J_A is large and positive, then the z components of the spin dominate in (1), and the system is described by an Ising model whereby the spins lie perpendicular to the magnetic plane; the $FeCl_2$-GIC system is believed to be an example of a 2D-Ising system. As the temperature and magnetic field are varied, unusual magnetic phase transitions are observed for the various types of spin alignments that can arise in these low dimensional systems. Of particular interest are the high stage compounds which exhibit 2D behavior and the stage 1 magnetic GICs, where cross-over behavior between 3D and 2D spatial dimensionality dominates.

The number of magnetic GICs that have been synthesized to date is small, and most of these have relatively low magnetic transition temperatures. Both donor (Eu) and acceptor ($FeCl_3$, $FeCl_2$, $CoCl_2$, $NiCl_2$, $MnCl_2$, $MoCl_5$) magnetic GICs have been synthesized and investigated with regard to their magnetic properties. The magnetic transition temperatures in the intercalation compounds tend to be significantly lower than in the pristine parent materials because the number of magnetic nearest neighbors is reduced by the planar geometry; in the intercalation compound, the c-axis nearest neighbors are at much larger distances.

Bi-intercalation allows the preparation of a superlattice with a periodicity consisting of the layer sequence ...CMCM'C... where M and M' are two different guest species and C is a graphite layer. By selection of M as a magnetic species and M' as non-magnetic, larger separations of the magnetic planes can be accomplished. Magnetic studies with such bi-intercalation samples are just beginning, and interesting results are expected.

3. Experimental Evidence for 2D-XY Behavior in GICs

Of the various possible types of 2D magnetic systems, the 2D-XY magnetic systems have been most widely studied, because of the special role these systems play in theoretical studies of critical phenomena, particularly with regard to their unusual magnetic phase transitions. The absence of long range order in 2D-XY systems [11] which show a magnetic phase transition was reconciled by KOSTERLITZ and THOULESS [12] who introduced the idea of spin vortices which form bound vortex pairs at the phase transition.

If the interaction between these spin vortices within the two-dimensional plane is governed by a (1/r) force, then the energy of a single vortex has a logarithmic decay

$$E = 2\pi J \ln(L/a), \qquad (2)$$

where J is the exchange energy, a is the nearest neighbor distance, and L is the sample size. Kosterlitz and Thouless showed that if a pair of vortices of opposite vorticity are brought together, they will bind at a temperature $T_{KT} \sim \pi J/k_B$ to produce a lower energy state with only short range correlations. For an ideal 2D-XY system, the magnetic susceptibility is expected to diverge at the Kosterlitz-Thouless transition temperature T_{KT}, following the functional form

$$\chi \sim \exp(b/t^{1/2}), \quad t > 0 \qquad (3)$$

where b is a constant and t is the reduced temperature

$$t = (T - T_{KT})/T_{KT} \qquad (4)$$

for $T > T_{KT}$ in the free vortex phase.

Experimental verification of the Kosterlitz-Thouless predictions requires the synthesis of a truly isolated magnetic layer of infinite extent and in zero magnetic field. The difficulty with the experimental realization of such a system has led theorists to extend the Kosterlitz-Thouless concept to finite size systems [13] and to systems with in-plane symmetry breaking fields [14] associated with crystalline ordering in the magnetic planes or externally applied fields.

From the experimental side, magnetic graphite intercalation compounds have been proposed as experimental systems which approximate 2D-XY behavior. To optimize the XY approximation, the intercalant is chosen to maximize the magnitude of (J_A/J) for a magnetic system where J_A and J have opposite signs, and to achieve 2D spatial behavior it is necessary that high stage compounds can be synthesized. Evidence is presented below that approximations to such magnetic phases can be achieved in magnetic GICs over a limited range of temperatures and magnetic fields.

Of the various magnetic GICs that have been synthesized, the $CoCl_2$-GICs exhibit the largest c-axis anisotropy ($J_A/J = 0.56$), and therefore have been selected by several research groups for the study of 2D-XY behavior. Furthermore, it is possible to synthesize $CoCl_2$-GICs for a variety of stages, and thus to investigate stage dependent effects and 2D-3D spatial cross-over phenomena. Unfortunately, the magnetic cobalt planes are not quasi-infinite (i.e., the magnetic planes do not extend throughout the sample), but rather break up into separated magnetic islands. However, the structural coherence lengths in the intercalate $CoCl_2$ islands are large enough to contain a statistically meaningful number of spins

($\sim 4 \times 10^3$ spins/island). This island structure is characteristic of the magnetic acceptor compounds having intercalate layers incommensurate to the adjacent graphite layers, such as for the $FeCl_2$, $FeCl_3$, $NiCl_2$, $MnCl_2$ intercalants.

The experimental work on magnetic GICs exhibiting 2D-XY behavior dates back to the early 1970s when Karimov and others first discovered anomalies in the magnetic susceptibility of a variety of transition metal chloride-GICs [15]. In this pioneering work, attention was also given to the identification of temperature ranges where spontaneous magnetization could be found and where anomalies in the specific heat could be correlated with anomalies in the susceptibility. For most acceptor 2D-XY systems that have been studied, two critical temperatures have been identified (T_{cl} and T_{cu}), thereby delineating three temperature regimes, each exhibiting distinct magnetic behaviors: (1) $T < T_{cl}$, (2) $T_{cl} < T < T_{cu}$, and (3) $T > T_{cu}$. At high temperature ($T \gg T_{cu}$), Curie-Weiss behavior is found [15]. Different magnetic field dependences are found at T_{cl} and T_{cu} for the various magnetic properties such as the magnetic susceptibility, magnetization and heat capacity, indicative of the different origins of these phase transitions. The magnetic phases are also observed to be stage dependent, with the behavior of the stage 1 $CoCl_2$-GICs distinctively different from that of the higher stage compounds (see Figure 2). It is the higher stage compounds that provide good approximations to 2D behavior.

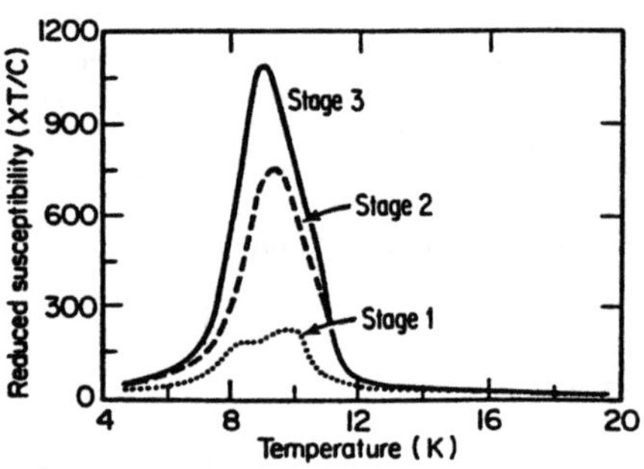

Fig. 2. Reduced susceptibility ($\chi T/C$) of stage 1, 2, and 3 $CoCl_2$-GICs vs. temperature. C is the Curie constant determined by high-temperature susceptibility measurements (C = 3.36 K emu/mole) [19].

Assuming that a Kosterlitz-Thouless transition occurs in high stage magnetic acceptor GICs, experimental studies indicate that the upper critical temperature T_{cu} corresponds to T_{KT} so that the region $T > T_{cu}$ is expected to exhibit free vortex behavior with the binding of the vortices occurring at T_{cu}. Because of the finite size of the 2D magnetic domains and other weak perturbations such as the in-plane symmetry-breaking 6-fold crystalline field anisotropy and the finite interplanar exchange coupling, the observed susceptibility for magnetic GICs does not become infinite at T_{KT}, but only becomes large. Therefore experimental studies have been directed to other evidence for 2D-XY behavior. Three measurements in particular support this point.

First, strong evidence that the high stage $CoCl_2$-GICs exhibit 2D-XY behavior comes from comparison of the experimental magnetic susceptibility for a stage 3 compound to the high temperature series expansion calculation (well above T_{cu}) for different spin dimensionalities [16,17]. Using previously determined values of C and J for pristine $CoCl_2$ [18], dimensionless plots of $\chi T/C$ vs. k_BT/J were calculated for the 2D-Heisenberg, 2D-XY, and 2D-Ising models, using no adjustable parameters, and the results are shown in Figure 3. A good fit to the experimental data is obtained for the 2D-XY model, while the fits to the 2D-Heisenberg and 2D-Ising models are poor.

Further evidence in support of the 2D-XY model comes from the extension of the Kosterlitz-Thouless calculations to include (1) finite size effects imposed by the observation of magnetic intercalate islands, (2) 6-fold in-plane crystalline symmetry-breaking fields, and (3) weak interplanar coupling between spins on

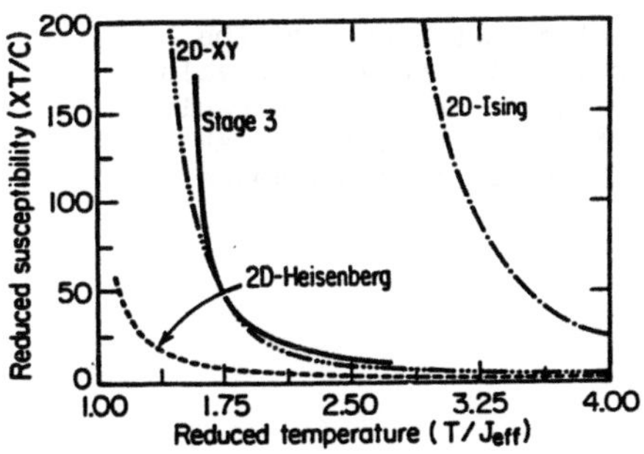

Fig. 3. Comparison of experimental susceptibility data with high temperature series expansion results for the classical 2D-Heisenberg, 2D-XY, and 2D-Ising models. The value of J_{eff} is 7.125 K and C is 3.36 K emu/mole [19].

different magnetic planes [19]. The number of spins contained in a magnetic island was determined by comparison of the calculated susceptibility and experimental measurements over a wide temperature range. From the comparison it was concluded that an average magnetic island in a stage 3 $CoCl_2$-GIC sample contained ~ 4000 spins, in good agreement with the direct observation of the island dimensions by transmission electron microscopy [20]. Further refinements are needed for the calculation to yield quantitative agreement with the calculated and measured temperature dependence of the susceptibility.

Inelastic neutron scattering measurements on a stage 2 $CoCl_2$-GIC further provide strong experimental evidence in support of the 2D-XY model [21]. In this experiment the spin configuration is probed directly. In addition to the magnetic Bragg peaks at (0,0,1/2) associated with the spin ordering below T_{cl} into an antiferromagnetic stacking of ferromagnetically ordered planes of spins, broad inelastic neutron scattering is observed. In these measurements, the integrated intensity of the Bragg scattering peak vanishes at T_{cl} while that for the broad diffuse background scattering falls rapidly as T_{cu} is approached (identified with the short range correlation of the vortex pairs) and a more gradual approach to zero scattering intensity is found above T_{cu} (identified with the much weaker scattering of the free vortices above T_{cu}). The availability of a theoretical model for the temperature dependence of the diffuse scattering for free and bound vortices would be especially helpful for a definitive identification of the binding of vortices at T_{cu}.

The neutron scattering studies show no dependence of the diffuse scattering intensity on wave vector along the c-direction in the temperature range $T_{cl} < T < T_{cu}$, consistent with the spins lying in the basal plane in a bound vortex configuration with strong in-plane spin-spin correlation but negligible interplanar correlation. Above T_{cu}, the absence of a remanent magnetization and a linear dependence of magnetization on magnetic field [16] identifies the phase above T_{cu} with paramagnetism, consistent with short range order of the spins in a free vortex configuration. The susceptibility near T_{cu} is very sensitive to the application of an external magnetic field and χ is rapidly suppressed by fields of a few Oe, as shown in Figure 4, as long range order is established [23,24]. A high sensitivity to magnetic field is expected for a Kosterlitz-Thouless transition, i.e., the bound vortex state is completely quenched in low fields [13].

An alternate interpretation of the observed magnetic properties in the temperature range $T_{cl} < T < T_{cu}$ and beyond has been proposed by RANCOURT [22] based on intraplanar and interplanar dipolar interactions between the magnetic islands. It is fair to say that at present, the experimental observation of a Kosterlitz-Thouless transition in magnetic GICs remains an open question.

4. 2D-3D Low Temperature Crossover in $CoCl_2$-GICs

Weak perturbations, such as the 6-fold in-plane symmetry-breaking crystal field H_6 and the interplanar magnetic interaction J' become

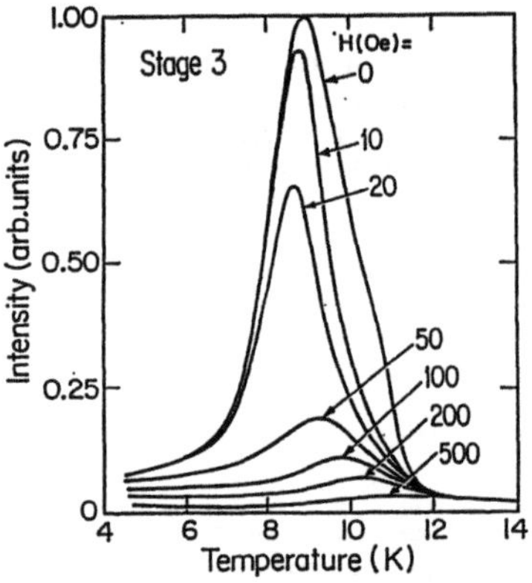

Fig. 4. Temperature dependence of the magnetic susceptibility for various values of the magnetic field, showing the rapid suppression of the susceptibility peak in stage 3 $CoCl_2$-GIC by an external magnetic field [24].

important at lower temperatures and give rise to a 2D-3D cross-over transition at T_{cl}. The spontaneous magnetization due to the coupling of the spins to H_6 produces ferromagnetically ordered in-plane domains below T_{cl}. At zero external magnetic field, the antiferromagnetic interplanar exchange interaction couples these ferromagnetic planar domains in an antiferromagnetic stacking arrangement, as implied by the Bragg peaks in neutron scattering experiments below T_{cl} [23]. Specifically, the (00ℓ) scan in the neutron scattering experiment clearly shows a half-order peak (ℓ = 1/2) associated with a magnetic doubling of the c-axis unit cell (below T_{cl}).

Since the magnitudes of these magnetic perturbations are very small, weak magnetic fields have a major effect on the magnetic properties of the $CoCl_2$-GICs. At lower temperatures, small externally applied magnetic fields compete with the small 6-fold in-plane anisotropy field in establishing the spin direction of the magnetic domains. The magnetic phase transitions observed upon application of a weak magnetic field in the plane of the spins, provide a convenient method for determination of J', which is the parameter most closely associated with the spatial dimensionality.

We now briefly discuss the interesting magnetic phases observed at low temperature in the stage 1 $CoCl_2$-GIC as a function of magnetic field. Assuming that the spins in a given magnetic plane are ferromagnetically aligned along an easy in-plane axis, with the antiferromagnetic interplanar exchange interaction strong enough to produce an antiferromagnetic arrangement of the ferromagnetically aligned planes, theoretical calculations predicted two phase transitions as a function of magnetic field [16,17]. At the lower critical field, a transition at H_{AS} from an antiferromagnetic to a spin-flop arrangement of the ferromagnetic planes was predicted, followed by a second transition at H_{SF} to a ferromagnetic interplanar alignment of the spins in the ferromagnetic planes along the magnetic field direction. Experiments subsequently carried out on the magnetic field dependence of the magnetic susceptibility at constant temperature for a stage 1 $CoCl_2$-GIC exhibited two peaks for temperatures below T_{cl} (see Figure 5). Based on these peaks, a magnetic phase diagram was constructed, as shown in Figure 6. In the absence of sufficient interplanar coupling J', the magnetization in the ferromagnetic planes would be aligned along any one of the six easy axis orientations with equal probability, and only a single broad peak would be expected, associated with the flipping of the magnetization from any one of the easy axis directions to alignment along the applied field direction. A single broad susceptibility peak is observed experimentally as a function of field for $CoCl_2$-GICs with stages $\geq$ 2 [24]. Magnetic field dependent behavior qualitatively similar to that for the stage 1 $CoCl_2$-GIC has also been reported for stage 2 $NiCl_2$-GICs [25].

A Monte Carlo simulation has been used to identify the spin configurations for the various magnetic phases occurring at low temperature as a function of magnetic

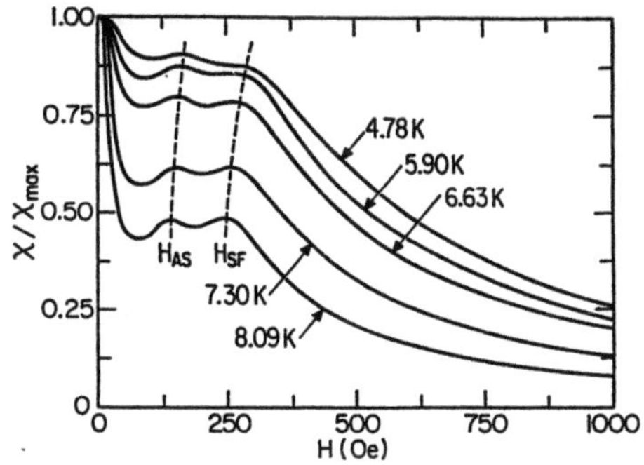

Fig. 5. Magnetic field ($\vec{H} \perp \vec{c}$-axis) dependence of the magnetic susceptibility for various values of temperature for a stage 1 $CoCl_2$-GIC. The two peaks in the susceptibility at H_{AS} and H_{SF} are identified with magnetic phase transitions associated with interplanar magnetic domain alignment [16,17].

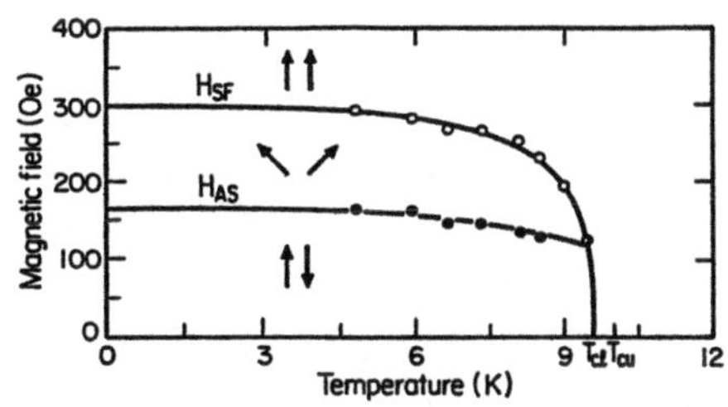

Fig. 6. Magnetic phase diagram for a stage 1 $CoCl_2$-GIC derived from the peaks in the magnetic susceptibility shown in Figure 5 [16].

field, and to evaluate the interplanar exchange coupling constant J' and the six-fold in-plane symmetry breaking field H_6 by fitting to the observed magnetic field spin-flop transitions. The Monte Carlo simulation is based on the Hamiltonian

$$H = J' \sum_{nn}(S_i^x S_j^x + S_i^y S_j^y) - g \mu_B S H_6 \sum_i \cos(6\theta_i) - g \mu_B \sum_i \vec{S}_i \cdot \vec{H} \qquad (5)$$

in which the main in-plane magnetic interactions J and J_A do not appear. The first term in (5) represents the interplanar antiferromagnetic exchange coupling between spins i and j on adjacent layers at the minimum interlayer spin-spin separation (the sum is over nearest neighbors nn), and the interplanar coupling constant J' is treated as a fitting parameter. The second term in (5) arises from the coupling between each spin and the intraplanar six-fold symmetry breaking field $\vec{H}_6$. Here θ_i is the spin angle with respect to the x-axis (the easy axis). The in-plane six-fold symmetry breaking field in the intercalation compound is almost identical to that in pristine $CoCl_2$ [18] and has a magnitude of ~10 Oe. The third term in the Hamiltonian (5) arises from the interaction between the spins and the applied external magnetic field. The g factor has a value of 5.4 and the effective spin value S is 1/2. In the Monte Carlo simulation, various orientations of the external magnetic field relative to the in-plane easy (x) axes were considered [24]. At very low temperatures the spins on a given layer are considered to be ferromagnetically aligned with respect to each other, and are represented by a single large classical spin.

The in-plane orientations of an array of these giant spins along the c-axis (vertical direction) obtained with the Monte Carlo calculations are plotted in Figure 7. Sequential columns represent the spin configurations at different values of magnetic field, sequential field

values differing by ΔH = 20 Oe. The results of Figure 7 confirm the magnetic phase assignments given above. The discontinuities in the average magnetization $\langle M \rangle$ and the hysteresis behavior at $H_{c2} = H_{SF}$ indicate that this transition is clearly first order. In contrast, the hysteresis effeect at $H_{c1} = H_{AS}$ is very weak. From the values of H_{c1} and H_{c2} in the low temperature limit, the value of J' can be found. Because of the small magnitude of H_6 in pristine $CoCl_2$, the values of H_{c1} and H_{c2} are not sensitive to H_6. The Monte Carlo study of the low temperature behavior provides important information on J' which is the dominant parameter for

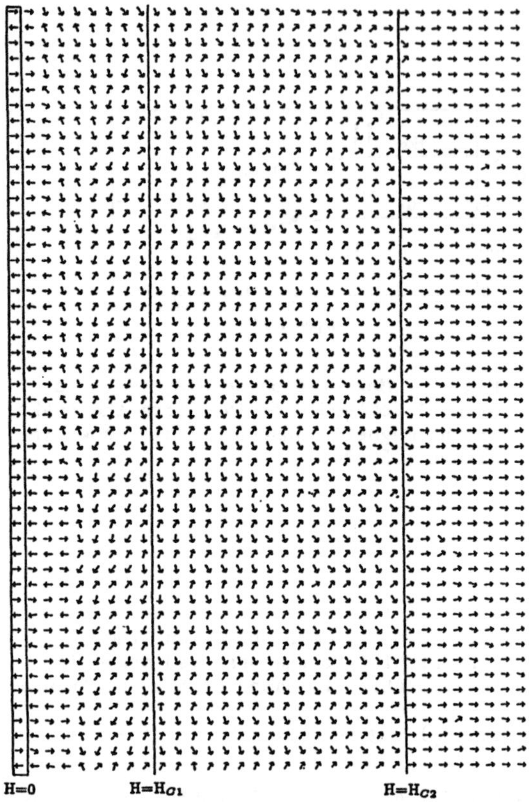

Fig. 7. Spin configurations in 20 Oe increments in the low temperature limit for the stage 1 $CoCl_2$-GIC [24].

determining the spatial dimensionality of the system, and hence the magnetic behavior near T_{cu}.

5. Concluding Remarks

The most serious factor impeding progress in the study of 2D magnetism in magnetic GICs is the synthesis of well-staged magnetic GICs of the appropriate intercalants and stages, and hence large in-plane coherence distances. In addition, further theoretical developments will be needed to treat a variety of perturbations which are present in magnetic GICs, including multiple symmetry-breaking fields, finite size effects, and weak interplanar coupling. Although these problems are very difficult, significant progress has been made for the case of the 2D-XY model with ferromagnetic nearest neighbor interactions. But many unanswereed questions remain concerning the spin behavior at the phase transitions in magnetic GICs.

Acknowledgments

The author wishes to thank Drs. G. Dresselhaus, S. T. Chen, and K. Y. Szeto for many important discussions. This work was supported by AFOSR Contract No. F49620-83-C-0011.

References

1. M. S. Dresselhaus, Physics Today, March (1984), p. 60.
2. M. S. Dresselhaus and G. Dresselhaus, Adv. in Phys. 30, 139 (1981).
3. S. A. Solin, Adv. in Chem. Phys. 49, 455 (1982).
4. F. R. Gamble and T. H. Geballe, in Treatise on Solid State Chemistry, edited by N. B. Hannay, Plenum Press, New York (1976), p. 89.

5. S. A. Solin, J. Mole. Catal. 27, 291 (1984).
6. P. Aldebert, N. Baffier, J. J. Legendre, and J. Livage, Rev. Chim. Min. 19, 485 (1982).
7. G. Timp and M. S. Dresselhaus, J. Phys. C: Solid State Phys. 17, 2641 (1984).
8. C. L. Fu, A. J. Freeman, and T. Oguchi, Phys. Rev. Lett. 45, 57 (1984).
9. A. Hérold, in Physics and Chemistry of Materials with Layered Structures, vol. 6, edited by F. Lévy, Reidel, Dordrecht (1979), p. 323.
10. A. Erbil, A. R. Kortan, R. J. Birgeneau, and M. S. Dresselhaus, Phys. Rev. B28, 6329 (1983); S. G. J. Mochrie, A. R. Kortan, R. J. Birgeneau, and P. M. Horn, Phys. Rev. Lett. 53, 985 (1984).
11. N. D. Mermin and H. Wagner, Phys. Rev. Lett. 17, 1133 (1966).
12. J. M. Kosterlitz and D. J. Thouless, in Progress in Low Temperature Physics, vol. 7B, edited by D. F. Brewer (North-Holland, Amsterdam, 1978), p. 395.
13. K. Y. Szeto and G. Dresselhaus, Phys. Rev. B32, 3187 (1985).
14. J. V. José, L. P. Kadanoff, S. Kirkpatrick, and D. R. Nelson, Phys. Rev. B16, 1217 (1977).
15. Yu. S. Karimov, Zh. Eksp. Teor. Fiz. 68, 1539 (1975) [JETP Lett. 41, 772 (1976)].
16. K. Y. Szeto, S. T. Chen, and G. Dresselhaus, Phys. Rev. B33, 3543 (1986).
17. K. Y. Szeto, Ph.D. Thesis, Massachusetts Institute of Technology (unpublished) (1985).
18. M. K. Wilkinson, J. W. Cable, E. O. Wollan, and W. C. Koehler, Phys. Rev. 113, 497 (1959); M. J. Hutchings, J. Phys. C6, 3143 (1973).
19. K. Y. Szeto, S. T. Chen, and G. Dresselhaus, Phys. Rev. B32, 4628 (1985).
20. S. T. Chen, K. Y. Szeto, M. Elahy, and G. Dresselhaus, J. Chim. Phys. 81, 863 (1984).
21. M. Suzuki and H. Zabel, Synth. Metals C12, (1985).
22. D. G. Rancourt, J. Mag. Mag. Mat. 51, 133 (1985).
23. M. Suzuki, H. Ikeda, and Y. Endoh, Synth. Metals C8, 43 (1983).
24. S. T. Chen, Ph.D. Thesis, Massachusetts Institute of Technology (unpublished) (1985).
25. M. Suzuki and H. Ikeda, J. Phys. C14, L933 (1981).

Structural and Magnetic Instabilities in Low-Dimensional Systems

F. Yndurain

Departamento de Física Fundamental, Universidad Autónoma de Madrid, E-28049 Madrid, Spain

We study the interplay between itinerant magnetic order and structural instabilities in low-dimensional systems. Mean-field calculations reveal that magnetic order can coexist with different kinds of charge and structural distortions like charge-density waves and Jahn-Teller instabilities. We find that in many cases the magnetic order induces a structural distortion which gives rise to a ferrimagnetic ground state. When electron correlation is included via second-order perturbation theory the structure of the mixed ground state remains unchanged, and a Kondo-like peak at the Fermi level appears.

1. Introduction.

In low-dimensional systems structural as well as charge-distribution instabilities are very common. Among them charge-density waves and Jahn-Teller distortions are the most frequently observed. In this work we study the interplay between these instabilities and magnetic order. As we will see, in some cases the electron-electron interaction suppresses the instability whereas in other cases it enhances it. Even more, the magnetic order produced by the Coulomb interaction can induce structural instabilities which give rise to a mixed and complex ground state.

To study the effect of the electron-electron interaction in charge-density waves we follow the model of Balseiro et al. [1]. The electron band structure is introduced by means of a tight-binding hamiltonian H_e of the form

$$H_e = V \sum_{i,j} c^\dagger_i \, c_j \quad , \tag{1}$$

where $c_i^\dagger$ (c_i) creates (destroys) an electron at site i. The electron-electron interaction is introduced via a Hubbard-like term H_{e-e} such that

$$H_{e-e} = U \sum_i n_{i\uparrow} \, n_{i\downarrow} \quad , \tag{2}$$

where $n_{i\uparrow}$ ($n_{i\downarrow}$) represents the number of electrons at site i with up (down) spin. This term is indeed responsible for the magnetic ordering. The lattice instability is represented by a Frölich-like electron-phonon Hamiltonian H_{e-p} of the form

$$H_{e-p} = - \tau \sum_k \; c^\dagger_{k+q} \, c_k \; , \tag{3}$$

where τ represents the lattice distortion parameter u multiplied by the electron-phonon matrix element g(q). The vector q is half a reciprocal lattice vector. Finally, the elastic energy is represented by the Hamiltonian

$$H_E = A \, \tau^2 \quad , \tag{4}$$

where A is the elastic constant divided by g^2.

The complete model Hamiltonian contains therefore four terms

$$H = H_e + H_{e-e} + H_{e-p} + H_E \; . \tag{5}$$

The solution of this Hamiltonian is very complicated in general. In the following section we will show how one can obtain solutions of (5) in the Hartree-Fock approximation for simple band structures such that the Fermi surface contains the topological characteristics of the charge density-wave formation

2. Mean-Field Solutions.

To simplify the Hamiltonian (5) we only retain the most important terms of (2) which is therefore simplified in the form

$$H_{e-e} = U \sum_{k,k',q}(n_{k\uparrow}\, n_{k'\downarrow} + c^{\dagger}_{k+q\uparrow}\, c_{k\uparrow}\, c^{\dagger}_{k'+q\downarrow}\, c_{k'\downarrow}) \quad . \tag{6}$$

The Hamiltonian (5) can now be treated in the Hartree-Fock approximation. It takes the form [1]

$$H_{H-F} = \sum_{k\sigma}(E_{k\sigma}\, c^{\dagger}_{k\sigma}\, c_{k\sigma} - \gamma_{\sigma}\, c^{\dagger}_{k+q\sigma}\, c_{k\sigma}) + A\, \tau^2 - U\, b_{\uparrow}\, b_{\downarrow} - U\, n_{\uparrow}\, n_{\downarrow} \; , \tag{7}$$

where

$$E_{k\sigma} = E_k + U\, n_{-\sigma} \quad , \tag{8}$$

$$n_{\sigma} = \sum_k \langle c^{\dagger}_{k\sigma}\, c_{k\sigma} \rangle \quad , \tag{9}$$

$$b_{\sigma} = \sum_k \langle c^{\dagger}_{k+q\sigma}\, c_{k\sigma} \rangle \quad , \tag{10}$$

$$\gamma_{\sigma} = \tau - U\, b_{\sigma} \; , \tag{11}$$

There are three relevant parameters in the model, namely the mean number of electrons $n = n_{\uparrow} + n_{\downarrow}$, the electron-electron Coulomb repulsion U and the lattice strength constant A. Depending on the values of these parameters the ground state of the system can be:

(i) A paramagnetic nondistorted state (P): $\tau = 0$ and $n_{\uparrow} = n_{\downarrow}$.
(ii) A pure charge-density wave state (CDW): $\tau \neq 0$ and $n_{\uparrow} = n_{\downarrow} = n/2$.
(iii) A pure ferromagnetic state (F) with $\tau = 0$ and $n_{\uparrow} \neq n_{\downarrow}$.
(iv) A pure spin-density wave state (SDW) with $\tau = 0$ ($b_{\uparrow} = -b_{\downarrow} \neq 0$) and $n_{\uparrow} = n_{\downarrow} = n/2$.
(v) A distorted state with some kind of magnetic order.

A. Quasi-One-Dimensional Model.

In this section we will discuss the model presented by Balseiro et al. [1]. The model one-electron Hamiltonian H_e is such that $E_k = -E_{k+q}$ such that the Fermi surface for the half-filled band has perfect nesting. The one electron energies are perturbed to take the values

$$E_{1k\sigma} = -\, U\, m_{\sigma} + (E_k{}^2 + \gamma_{\sigma}{}^2)^{1/2} \; , \tag{12}$$

$$E_{2k\sigma} = -\, U\, m_{\sigma} - (E_k{}^2 + \gamma_{\sigma}{}^2)^{1/2} \; , \tag{13}$$

where m_{σ} is the magnetization (with a plus or a minus sign) and

$$\gamma_{\sigma} = (1/2A)(b_{\uparrow} + b_{\downarrow}) - U\, b_{-\sigma} \; . \tag{14}$$

The self-consistent equations to solve are:

$$n_{\sigma} = (1/2) \sum_k [f(E_{1k\sigma}) + f(E_{2k\sigma})] \tag{15}$$

and

$$b_{\sigma} = -\gamma_{\sigma} \sum_k (1/2)\, (E_k{}^2 + \gamma_{\sigma}{}^2)^{-1/2}\, [f(E_{1k\sigma}) - f(E_{2k\sigma})] \quad , \tag{16}$$

where f is the Fermi distribution function. Equations (15) and (16) can be solved self-consistently to give the stable solutions of the Hamiltonian. To chose the

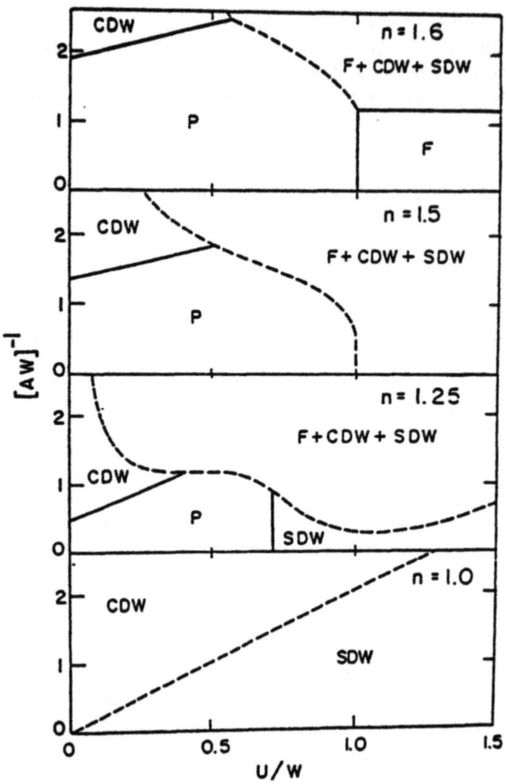

Figure 1. Phase diagram $(AW)^{-1}$ vs. U/W (W is the bandwidth) for different occupation numbers n [1]. The broken lines represent transitions where the order parameters are discontinuous at the transitions and full lines represent transitions where the order parameters are continuous.

more stable one (ground state) we pick up the solution with lowest energy. The phase diagram at T = 0 for different electron occupations is shown in Figure 1.

For n = 1 (i.e. perfect nesting) the ground state is either a CDW or a SDW depending [2,3] on the values of the parameters U and A. However, for $n \neq 1$ different broken symmetries can coexist. In particular the mixed state F+CDW+SDW occupies an important portion of the phase diagram. This mixed ground state is such that without ferromagnetism the Fermi level lies far from the nesting condition, but if the bands are shifted by the magnetic interaction either the minority or majority spin band can meet the nesting condition and lowers the energy of the system creating a CDW and opening a gap in the spectrum (see Figure 2). It is easy to see that the simultaneous presence of ferromagnetism and the charge-density wave induces a spin-density wave such that the ground state has ferrimagnetic order.

B. Two-Dimensional Band Structure With Saddle Points.

RICE and SCOTT [4] have shown that if the band structure presents saddle points near the Fermi level, the electron-phonon coupling can give rise to the formation of a CDW. Here the mechanism is different from the one proposed by Peierls since no gaps are formed in the spectrum. The solution of the model Hamiltonian (5) for a saddle point has been obtained by BAUSELLS and YNDURAIN [5] in the case of nesting and in the general case. The model unperturbed density of states is given by

$$N(\varepsilon) = \begin{cases} (1/W)\ \ell n\ [W/2|\varepsilon|] , & \text{if } |\varepsilon| < W/2 , \\ 0 , & \text{otherwise} . \end{cases} \tag{17}$$

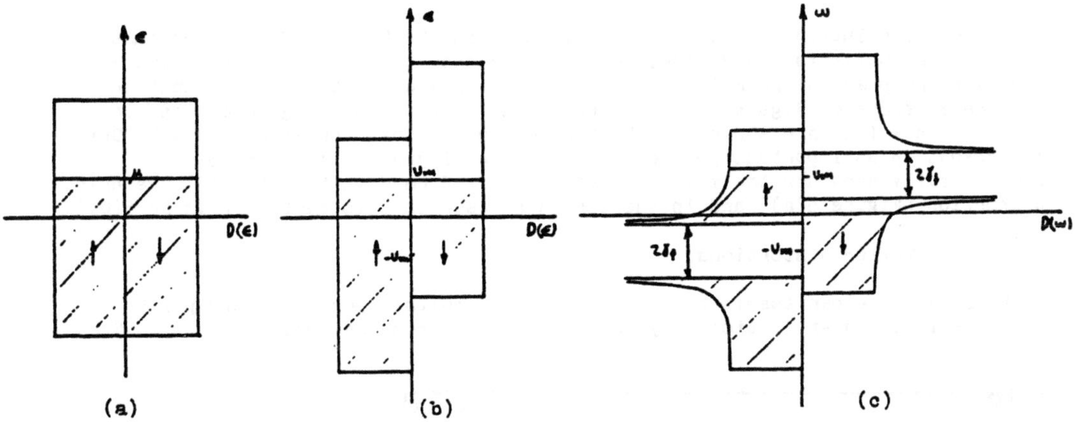

Figure 2. Mechanism of formation of the mixed ground state F+CDW+SDW. In (a) we represent the density of states in the undistorted paramagnetic phase (it is assumed that the perfect nesting condition occurs at $\varepsilon = 0$). In (b) the bands are shifted because of the appearance of ferromagnetic order in such a way that the minority-spin band meets the nesting condition. In (c) the electron-phonon interaction opens gaps in the two spin subbands giving rise to the stable configuration.

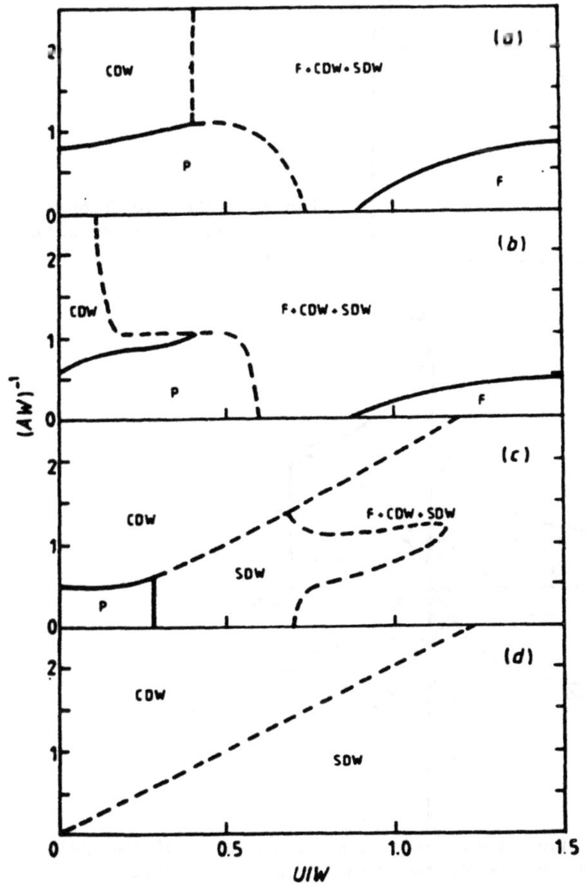

Figure 3. Phase diagram $(AW)^{-1}$ vs. U/W (W is the bandwidth) for different occupation numbers n when the two-dimensional band structure presents saddle points and there is nesting of the bands [5]. (a), (b), (c), and (d) correspond to n = 1.6, 1.5, 1.25, and 1, respectively. The broken lines represent transitions where the order parameters are discontinuous at the transitions, and full lines represent transitions where the order parameters are continuous.

Since the solution of the model Hamiltonian in the Hartree-Fock approximation follows the above-discussed line, we will only show the results. In Figure 3 we indicate the phase diagram in the case of nesting and for the same occupation numbers as those of Figure 1. In Figure 4 the results in the general case for n = 1 and n = 1.25 are shown. In both cases we see that the mixed state F+CDW+SDW is present in an important portion of the phase diagram. In Figure 5 we schematically show how the mixed ground state is formed in the case of a Peierls distortion [Figure 5(a)] and in the case discussed in this section [Figure 5(b)].

C. Jahn-Teller Distortions.

In this case one can assume that there are two orbitals per atom (two p-like orbitals in a planar lattice, say) and that the system has cubic symmetry. The

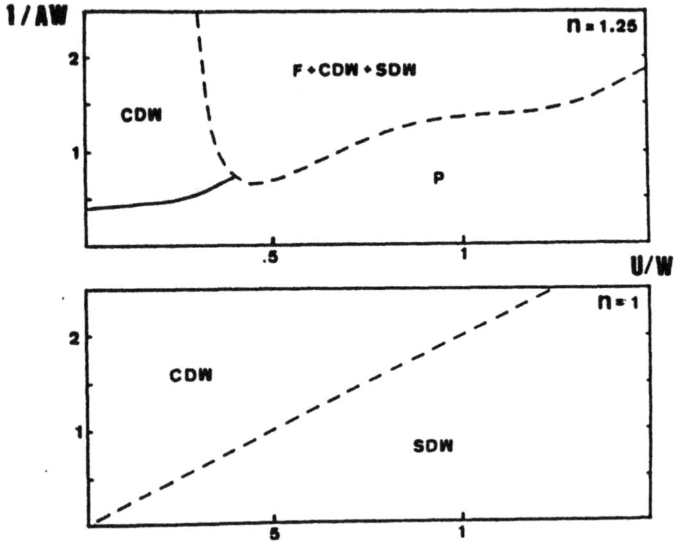

Figure 4. Same as Figure 3 for the general case without nesting.

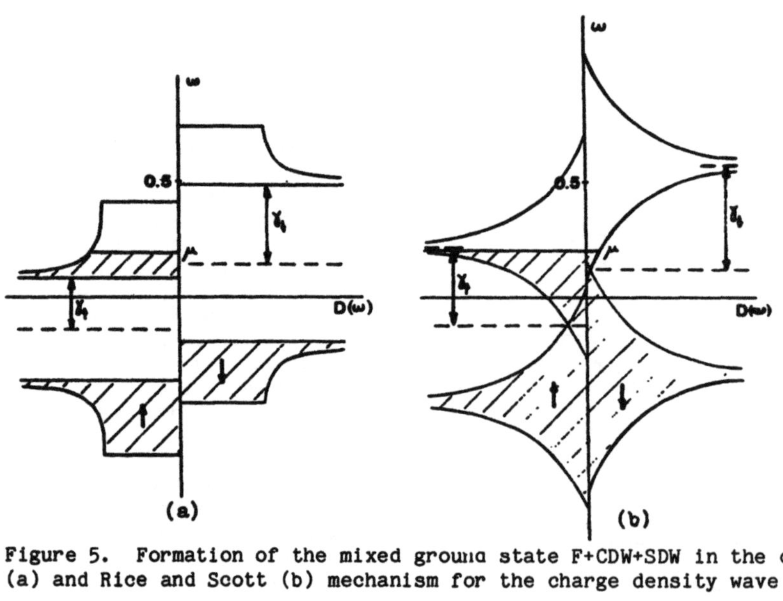

Figure 5. Formation of the mixed grouna state F+CDW+SDW in the case of the Peierls (a) and Rice and Scott (b) mechanism for the charge density wave formation.

model Hamiltonian can be written as before except for the electron-phonon term which in this case takes the form

$$H_{e-p} = -\tau \sum_{i,\sigma}(n_{i2\sigma} - n_{i1\sigma}) \quad , \tag{18}$$

where 1 and 2 stand for the two orbitals in each atom. The system lowers its energy if the two subbands are populated with a different number of electrons; this is accompanied by a tetragonal lattice distortion making the two atomic orbitals nonequivalent. The solution of the model Hamiltonian in the Hartree-Fock approximation it is easy if one assumes a simle density of states. In this case one obtains again the mixed ground state, i.e., a lattice distortion accompanied with magnetic order.

3. Correlation Effects.

Thus far we have only found solutions of the model Hamiltonian in the mean-field approximation. In this section we study the effect of the electronic correlation [7] in second-order perturbation theory in the electron-electron interaction parameter U.

We start with the following unperturbed band structure in a two-dimensional simple cubic lattice

$$E_k = 2V\,(\cos k_x a + \cos k_y a) \quad , \tag{19}$$

where a is the lattice constant.

We have solved the model Hamiltonian (5) in the Hartree-Fock approximation for $U = 3V$, $n = 1.25$, and $A = 1/(2V)$. Results of the calculation are shown in Figure 6(a). As we can see immediately, we get again a mixed ground state with a minority-spin band semiconducting and a majority one metallic. The analysis of the partial densities of states reveals that we have both a CDW and a SDW with a periodicity c(2x2) with two atoms per unit cell, such that the new unit cell is a square of half the area and rotated 45° with respect to the original one. The magnetic moment of one atom is 0.54 and points upwards, whereas the other atom in the unit cell has a moment 0.07 and points downwards. To include correlation in this structure, we follow the procedure of TREGLIA *et al.* [8]. In this case the calculation is more complicated than in the case of one band only, and several contributions to the self-energy have to be considered to account for the various second-order processes which take place. In addition to the spin-flip contributions to the self-energy, intra-atomic and inter-atomic excitations have to be considered.

Results of the final densities of states are shown in Figure 6(b). We immediately notice the shrinking of the bands due to the correlation effects as well as the tails beyond the original band edges. These results are the general features produced by the electronic correlation [8]. The analysis of the partial densities of states again shows that the ground state has both charge and spin broken symmetries and whereas the minority-spin band is semiconducting the majority one is metallic (the Fermi level lies at a continuum in the density of states). However, there is a new and important difference in the new results: the Fermi level in the majority-spin band lies at a narrow peak in the density of states. Although it is difficult to assess the origin of this feature given the numerical character of the calculation, we believe that it has a Kondo origin since we are dealing with a large magnetic moment (0.54) surrounded by nearest-neighbor atoms with a very small magnetic moment (-0.07). This new feature appears when electrons in nonpolarized atoms are correlated around a large magnetic moment [9].

4. Concluding Remarks.

The above discussion indicates that the mixed ground state with both charge and spin broken symmetries is likely to exist in low-dimensional systems. It has been

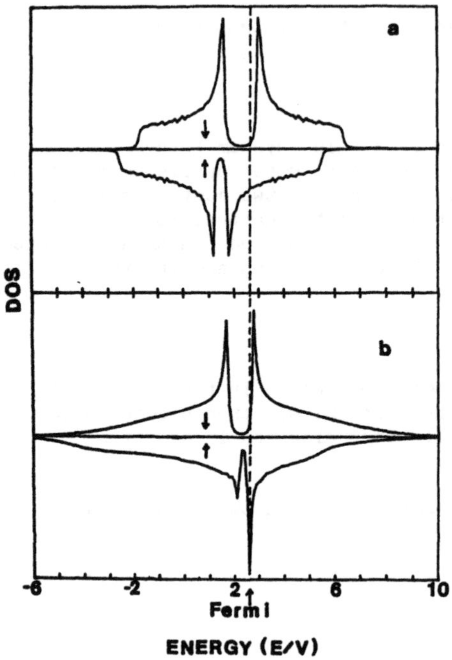

Figure 6. Effect of the electron correlation on the formation of the ground state F+CDW+SDW in a two-dimensional simple cubic lattice. In (a) the spin-resolved self-consistent densities of states in the Hartree-Fock approximation are presented. In (b) the final densities of states after the inclusion of the electron correlation in second order perturbation theory are shown. Notice the Kondo-like peak in the majority spin subband at the Fermi level. A small (0.05 V) contribution to the energy has been added for computational purposes.

postulated to take place in various situations, although its existence has never been confirmed. It was originally suggested [10] to be the ground state of monolayers of Co on top of Cu. KIVELSON and HEIM [11] considered the possibility of a mixed ground state in polyacetylene. DE GROOT et al. [12] considered the possibility of a mixed ground state to explain the giant magneto-optical Kerr rotation of PtMnSb. Finally, COLEMAN et al. [13] suggested that the strong magnetic field dependence of the resistance of $NbSe_3$ is due to a mixed ground state. All these works indicate that the existence of complex ground states as those discussed here is possible but their experimental characterization is very difficult for at present, since independent experimental probes have to be used.

References.

1. C. A. Balseiro, P. Schlottman, and F. Yndurain, Phys. Rev. B 21, 5267 (1980).
2. D. Dendley and L. M. Falicov, Phys. Rev. B 17, 1289 (1978), and S. K. Chan and V. Heine, J. Phys. F 3, 795 (1973).
3. D. Penn, Phys. Rev. 142, 350 (1966).
4. T. M. Rice and G. K. Scott, Phys. Rev. Lett. 35, 120 (1975).
5. J. Bausells and F. Yndurain, J. Phys. C 17, 5539 (1984).
6. S. K. Ghatak and D. K. Ray, Phys. Rev. B 31, 3064 (1985).
7. L. Milans and F. Yndurain, private communication and to be published.
8. G. Treglia, F. Ducastelle, and D. Spanjaard, J. Physique 41, 281 (1980).
9. C. Lacroix, J. Phys. F 11, 2389 (1981).
10. L. Gonzalez, R. Miranda, M. Salmeron, J. A. Verges, and F. Yndurain, Phys. Rev. B 24, 3245 (1981).
11. S. Kivelson and D. E. Heim, Phys. Rev. B 26, 4278 (1982).
12. R. A. de Groot, F. M. Mueller, P. G. van Engen, and K. H. J. Buschow, Phys. Rev. Lett. 50, 2024 (1983).
13. R. V. Coleman, G. Eiserman, M. P. Everson, A. Johnson, and L. M. Falicov, Phys. Rev. Lett. 55, 863 (1985).

Index of Contributors

Strong and Ultrastrong Magnetic Fields and Their Applications

Editor: F. Herlach

1985. 199 figures. XII, 362 pages. (Topics in Applied Physics, Volume 57). ISBN 3-540-13504-9

Contents: *F. Herlach:* Introduction. – *J. Hajdu, G. Landwehr:* Quantum Transport Phenomena in Semiconductors in High Magnetic Fields. – *M. Motokawa, L. W. Roeland:* High Field Magnetism. – *K. Dransfeld, G. Maret:* Biomolecules and Polymers in High Steady Magnetic Fields. – *D. B. Montgomery:* High Field Magnetic Confinement of Fusion Plasmas. – *F. Herlach, N. Miura:* Pulsed and Ultrastrong Magnetic Fields. – Editor's Appendix. – Subject Index.

This is the first comprehensive review of experiments with very strong magnetic fields above 10–15 T and up to several megagauss. These are only available at a few large *magnet laboratories* which are described in the introduction. The discussion begins with *magnetotransport* phenomena, mainly in semiconductors and in two-dimensional systems (quantum Hall effect) with an extensive treatment of the theoretical background. *Magnetism* is treated with a view to the discrimination between different interactions which may be sorted out by using very high fields, and magnetic resonances in the far infrared. The study of *biomolecules* in strong magnetic fields was developed recently and this is indeed the first review of this topic. Polymers, liquid crystals, biological membranes, the separation of macromolecules in strong field gradients, and the influence of magnetic fields on living organisms are treated. Advantages and trade-off considerations regarding the magnetic field strength in a *nuclear fusion* reactor are discussed with the example of the Alcator high field tokamak; this gives a good example of modern high field magnet design. Methods of generating *pulsed magnetic fields* up to 1000 T are described in practical detail and a variety of solid state experiments are presented with an outlook to further development.

Springer-Verlag
Berlin Heidelberg New York
London Paris Tokyo

MIX
Papier aus verantwortungsvollen Quellen
Paper from responsible sources
FSC® C105338

If you have any concerns about our products, you can contact us on
ProductSafety@springernature.com

In case Publisher is established outside the EU, the EU authorized representative is:
Springer Nature Customer Service Center GmbH
Europaplatz 3, 69115 Heidelberg, Germany

Printed by Libri Plureos GmbH
in Hamburg, Germany